In the footprints of the first naturalist explorers

250 years of biological exploration in Polynesia

Justin Gerlach

ISBN: 978-0-9932203-5-7

© Justin Gerlach, 2018

Phelsuma Press
Cambridge
UK

http://islandbiodiversity.com/phelsumapress.htm

Contents

Introduction	5
Tahiti	7
Mauipiti	41
Moorea	59
Huahine	63
Raiatea & Tahaa	79
Bora Bora	105
References	113

Introduction

I've had an interest in the islands of Polynesia for the past 26 years, since carrying out some of my PhD research there. In looking at historical documents to see if how much the islands had changed in recorded history I found that my most recent field studies happened to be exactly 250 years since the first European discovery and exploration of the islands. In 1767 the British ship HMS *Dolphin*, commanded by Samuel Wallis, landed on Tahiti. Thus begins the written record of the islands. The British claim to having 'discovered' the islands was narrowly achieved, with the *Etoile* and the *Boudeuse* under the great French explorer Louis-Antoine de Bougainville thinking the islands were their discovery less than 10 months later. Bougainville's expedition was notable because he was accompanied by the world's first expedition naturalist, Philibert de Commerçon.

The following account is a description of my exploration of the Society islands in 2017 and 1992, compared to the accounts made by explorers over the past 250 years. In that time the islands have experienced many changes, but in many respects it is remarkable how much has stayed the same.

The purpose of my visits to the islands has been snail conservation and research. While this may seem an unusual, if not simply eccentric reason it is not actually that strange. The *Partula* tree snails of the Society islands, Tahiti and Moorea in particular, have attracted more biologists and for longer than any other group of animal or plant in the islands. Again this goes back to the early explorers; this time the third expedition to visit the islands. Captain James Cook's arrival in the islands in 1679 on the *Endeavour* was notable for two reasons. Firstly it was part of the very first international scientific collaboration, the simultaneous observation of the 'Transit of Venus' by several different European expeditions. Secondly, it was the first British expedition to have a dedicated naturalist as national rivalries led the British to attempt to emulate Commerçon's remarkable discoveries. Joseph Banks was selected as the naturalist and brought with him a collecting team, including the meticulous Daniel Solander. Even though Banks has the fame, most of the collecting was done by Solander. This includes the first *Partula* snail to be brought to the attention of the scientific world.

Just over 200 years later the accumulation of masses of pretty *Partula* shells by European and American naturalists caught the attention of the pioneering evolutionary biologists. Serious scientific study started with the work of Henry Crampton in 1906 and carried on almost uninterrupted until the 1980s. The introduction of alien snails put an end to most of this work. First the giant African snail was imported to farm as 'escargots' then, when that was abandoned, and the snails released, the carnivorous *Euglandina* wolf snail was introduced to eat them. By 1974 when the *Euglandina* were introduced it was well known that they would not control the giant African snails, having little liking for such large, tough prey. Predictably they turned their attention to the inoffensive tree snails and wiped out dozens of species of *Partula* in the space of a few years.

My first visit to Polynesia, in 1992, was to study the ongoing invasion by *Euglandina* and to rescue the last surviving wild *Partula*. I spent only a little time on Tahiti and focussed my work on Raiatea island where there were still some *Partula* left. Zoos have been breeding the snails that I, and others, rescued with the hope of being able to reintroduce them to the wild one day. For the past 23 years the situation in the islands has been monitored by Trevor Coote. Trevor found that there were a few tiny populations of *Partula* that still survive even with the predators. He also found that *Euglandina* had become rare in many places, perhaps having eaten its way through all the available food. This suggested that reintroductions might be practical and accordingly this was attempted in 2015 and subsequently.

I wanted to see the progress of the reintroductions and the state of the wild *Partula* populations. So my 2017 visit was planned to cover these, and to search for other populations that might exist in rarely visited places. To do this I planned an itinerary that would take me to all of the high Society islands. Historically all landfall in the islands starts on Tahiti.

Tahiti

"we discovered... a long low island, with a white beach, of a pleasant appearance, full of cocoa-nut and other trees, and surrounded with a rock of red coral. We stood along the north east side of it, within half a mile of the shore; and the savages, as soon as they saw us, made great fires, as we supposed, to alarm the distant inhabitants of the island, and ran along the beach, abreast of the ship, in great numbers, armed in the same manner as the natives of the Islands of Disappointment. Over the land on this side of the island we could see a large lake of salt water, or lagoon, which appeared to be two or three leagues wide, and to reach within a small distance of the opposite shore... At this place the natives have built a little town, under the shade of a fine grove of cocoa-nut trees."

This first reported sighting of Tahiti (Matavai point) by Samuel Wallis on HMS *Dolphin* on 24th June 1767 is markedly different from the modern traveller's first sight on Tahiti. These days almost everyone arrives by air, more often than not arriving at night. The disorientated traveller steps off the aeroplane to be welcomed by a small troupe of Tahitian singers before entering what could be one of hundreds of other small airports.

View of Tahitit's Faa'a airport, shortly after arrival

From the airport you could take another disorientating flight straight to another island, move to a fancy hotel somewhere round the coast or travel into the town of Papeete. This is what I did on my arrival in August 2017, just over 250 years since Wallis.

I walked into Papeete just as the sun rose, getting an impression of the town as it emerged from the darkness. The roads were quiet at this time although very busy later on. The only animals apparent were the widely introduced Indian mynah birds and red-vented bulbuls, both most conspicuous at dawn.

The coastal road gave glimpses of the extraordinary appearance of Moorea once the sun was up. With its steep peaks Moorea looks more like an impression of a volcanic island rather than something real.

Dawn rises over Tahiti (top) and over the view of Moorea (bottom).

Herman Melville described Papeete in 1842 in 'Omoo': "The village of Papeetee [sic] struck us all very pleasantly. Lying in a semicircle round the bay, the tasteful mansions of the chiefs and foreign residents impart an air of tropical elegance, heightened by the palm-trees waving here and there, and the deep green groves of the Bread-Fruit in the background. The squalid huts of the common people are out of sight, and there is nothing to mar the prospect. All round the water, extends a wide, smooth beach of mixed pebbles and fragments of coral. This forms the thoroughfare of the village; the handsomest houses all facing it the fluctuations of the tides being so inconsiderable, that they cause no inconvenience."

Papeete may have had charm long ago, but it is hard to see now. When I first went there in 1992 it was very run down. The buildings were unprepossessing with the exception of the recently built market. This was an oasis of colour, light and industry in an otherwise depressed town. There were a few smart and vastly overpriced cafes and restaurants on the main road, occupied by rich French tourists. In the second row of streets there were more basic fast food outlets that attracted Australian back-packers, and other than that, almost everything was closed. Today the town seems much more affluent, with more shops and more activity. The market

Rue Rivoli in 1900 (photo L. Gauthier)

Papeete main street in 1951, and today

is still the same, and still the bright point of Papeete. It is densely crammed with stalls packed with vegetables and fruit, or with fish, or herbs, spices and oils, or with a remarkable profusion of flowers.

Papeete market

Shell necklaces in Papeete market

The restaurants are still fantastically expensive but have a more varied clientele than before. The bright spot of dining in the Society islands has always been the roulettes, the food-trucks that can be found in parking lots or on waste ground. Twenty five years ago they were exactly that; just a van with an open side and a shelf, serving pre-prepared or easily prepared food: Polynesian snacks, some Chinese dishes, sashimi or Polynesian raw fish in coconut milk. Now another level of sophistication has been added, with tables and chairs and each one has at least one portable grill. The food is always good and although the number of dishes is limited they are geographically varied. There is the staple of the French colonies, 'steak frites', although this being Polynesia the traditional steak and chips can be expanded with the addition of a generous portion of rice. There is an Asian influence, usually in the form of chow mein, with individual portions that look big enough to feed a family. Finally there is the Polynesian traditional raw fish in coconut milk, varying from the crude to the refined, but always good. At a reasonable price and with vast portions the roulettes are always very busy. With 60% of the population being obese, it is not surprising to see the size of the portions or to observe that everyone is always eating.

The greater prosperity of the islands compared to twenty five years ago is reflected in a change in demeanour of the population. In 1992 I was struck by how well Gauguin had captured what seemed to be the local expression. It isn't a uniquely Polynesian expression, it is something that can be seen in many islands around the world. Perhaps it is common to all islands; an easy slip into tedium. It isn't such a common expression now, perhaps the tedium of island life is staved off by the near universality of smart phones, or the possibility of being able to occupy yourself with music even when away from home, either by turning up the volume before you leave or by carrying it with you. Whatever the cause, the blank ennui isn't as obvious as it once was.

The main settlement before the establishment of the modern city of Papeete was a short distance to the east, at Matavai point. Wallis's described the Matavai settlement (what he called Port Royal): "We saw many houses or wigwams of the natives, but they were totally deserted, except by the dogs, who kept an incessant howling from the time we came on shore till we returned to the ship: they were low mean hovels, thatched with cocoa-nut branches; but they were most delightfully situated in a fine grove of stately trees, many of which were the cocoa-nut, and many such as

Matavai village by Duperry (1823) on the voyage of the *Coquille*

Tahitian house in 1910 from Christian (1910)

we were utterly unacquainted with. ... We observed the shore to be covered with coral, and the shells of very large pearl oysters; so that I make no doubt but that as profitable a pearl fishery might be established here as any in the world." Wallis was right about the oysters, as today pearls are one of the main exports from the islands and adverts for Polynesia's distinctive black pearls are everywhere.

Modern Tahiti still has the abundant dogs, but the houses are different. The wigwams have long been replaced by block houses roofed with corrugated iron sheets, indistinguishable from houses around most of the world.

Captain James Cook made Matavai Bay his base of operations during his first voyage to the Pacific. This Transit of Venus expedition is the most famous of all the expeditions to Tahiti. Cook's ship, the *Endeavour*, reached Matavai on 3rd June 1768. He established a shore camp there and set up an observatory at Mahina point. Over the next two months his task was to prepare to observe the movement of Venus across the face of the sun. This effort to calculate the distance between the sun and the earth using the rate of the transit was the first international scientific collaboration and marks the start of the modern scientific age. The sand spit at the end of Mahina point is also known as Point Venus. Although it is perhaps historically the most important spot in Polynesia, Point Venus is not much to look at. Today most of it is a suburb with a small park area. The park was very run down for many years but is now well kept. In the centre of it is a stone commemorating the visit of the *Bounty* in 1788 and a small octagonal lighthouse. This was built in 1860s, combining the skills of French designers, builders and lens manufacturers. Robert Louis Stevenson claimed that it had been designed by his father, but this has no basis in fact.

Papeete was just a coastal village until the end of the 18th century when its harbour expanded for the security it offered to the increasing numbers of ships visiting the island. By the 1820s Papeete had developed into a proper port and had become Queen Pomare IV's capital. As described by the US Exploring Expedition "The scenery at Papieti is remarkable; the background is filled up with a number of pinnacle-shaped mountains, jutting up in a great variety of forms; beneath, and directly in front of them, lies the semi-circular harbour, surrounded by the white cottages and churches of the village, embosomed in luxuriant foliage... In front, the little oral island of Moto-utu forms an embellished foreground, and serves to break the regularity of the line of the harbour, while by concealing its extent, it gives it an air of greater magnitude than it in reality possesses."

Herman Melville described Motu Utu "Right in the middle of Papeetee harbor is a bright, green island, one circular grove of waving palms, and scarcely a hundred yards across. It is of coral formation; and all round, for many rods out, the bay is so shallow, that you might wade any where. Down in these waters, as transparent as air, you see coral plants of every hue and shape imaginable: antlers, tufts of azure, waving reeds like stalks of grain, and pale green buds and mosses. In some places, you look through prickly branches down to a snow-white floor of sand, sprouting with flinty bulbs; and crawling among these are strange shapes: some bristling with spikes, others clad in shining coats-of-mail, and here and there, round forms all spangled with eyes.

Fort Venus by S. Middiman in 1773 (in Parkinson 1784)

"The island is called Motoo-Otoo; and around Motoo-Otoo have I often paddled of a white moonlight night, pausing now and then to admire the marine gardens beneath. The place is the private property of the queen, who has a residence there a melancholy-looking range of bamboo houses neglected and falling to decay among the trees."

Motu Utu is now joined to Tahiti having been incorporated into the harbour in 1962.

The one thing that is special about Papeete is the view inland. As Herman Melville wrote of Tahiti: "Seen from the sea, the prospect is magnificent. It is one mass of shaded tints of green, from beach to mountain top; endlessly diversified with valleys, ridges, glens, and cascades. Over the ridges, here and there, the loftier peaks fling their shadows, and far down the valleys. At the head of these, the waterfalls flash out into the sunlight as if pouring through vertical bowers of verdure. Such enchantment, too, breathes over the whole, that it seems a fairy world, all fresh and blooming from the hand of the Creator.

"Upon a near approach, the picture loses not its attractions. It is no exaggeration to say, that to a European of any sensibility, who, for the first time, wanders back into these valleys away from the haunts of the natives -the ineffable repose and beauty of the landscape is such, that every object strikes him like something seen in a dream; and for a time he almost refuses to believe that scenes like these should have a common place existence. No wonder that the French bestowed upon the island the appellation of the New Cytherea."

Not everyone was as impressed; an account of the US Exploring expedition in 1839 commented "The beauty of the distant view of Tahiti has been celebrated by all navigators, but I must confess that it disappointed me. The entire outline of the island was visible for too short a time and at too great a distance to permit its boasted features to be distinctly seen. Upon a second and nearer view, its jagged peaks and rugged inaccessible mountains were visible, but we looked in vain for the verdant groves which are said by all writers to clothe it. These indeed exist, but are confined to a narrow belt of low land, lying between the mountains and the shore, and being unseen at a distance, the general aspect of the island is that of a land recently thrown up by volcanic action." On closer inspection he modified his description: "Even upon the steep surface of its cliffs, vegetation abounds; the belt of low land is covered with the tropical trees peculiar to Polynesia; while the high peaks and wall-faced mountains in the rear are covered with vines and creeping plants... ridges diverge to all parts of the coast, throwing off spurs as they descend. These ridges are precipitous, and for the most part narrow. In many instances their summit is a mere edge, making walking upon them not only dangerous, but often impossible."

When the clouds clear the view from Papeete is fantastic, with the mountains rising up dramatically behind the coastal road. In 1767, the captain of the second expedition to reach Tahiti, the French Louis-Antoine Comte de Bougainville, described "The view of the coastline around us, elevated like an enormous amphitheatre, was a magnificent spectacle to behold. Although the mountains here are of a great height none of the outcropping rock gives any impression of barrenness. Nowhere was there naked rock to be seen; everywhere everything was clothed in a mantle of trees.

"We could hardly believe our eyes when we saw that one of the highest peaks was completely covered in trees right up to its very topmost isolated point. This peak was of an elevation equal to

Tahiti: 1849 by E.G. Fanshawe (top: National Maritime Museum, Greenwich, London) and today (below)

17

George Tobin exploring Matavai valley in 1792 on Bligh's second expedition (National Library of Australia)

that of the mountains to be seen in the interior at the southern-most part of the island. This pinnacle appeared to have a diameter of no more than 30 fathoms and the higher it climbed the narrower it became. If one looks at it from afar one would have taken it for a pyramid of immense height that had been dressed with garlands and strands of foliage by the hand of some skilled superhuman florist. The less elevated regions of the countryside are interspersed with areas of meadows and stands of timber.

"Along the entire length of the coastline, extending back as far as the commencement of the more elevated land there is a border of low-lying and level land covered with plantations. It is here amongst banana, coconut and other fruit bearing trees that the habitations of the islanders are located."

Bougainville was rapturous about the countryside and was far from objective: "On several occasions, in the company of the second or third officer, I took a long walk into the interior of the island. It always seemed to me that I had been transported into the Garden of Eden. We would wander through meadows filled with beautiful fruit trees amongst which meandered small streams the sparkling waters which imparted upon the atmosphere a delicious coolness and freshness without provoking even the slightest inconvenient increase in humidity.

"In this well-populated paradise the fortunate inhabitants have only to languorously stretch out a hand in order that they may enjoy the benefits and blessings of green and munificent nature. Often we came across groups of men and women sitting together in their shady orchards; they would call out to us from their grassy couches cheerful and friendly greetings. Whenever we met people walking along the pathways they would courteously step to one side that we might pass without inconvenience. Wherever and everywhere we looked we saw hospitality, relaxation, joy in simply being alive and every possible outward expression of happiness." Although the basis of his report is probably accurate his description is based at least in part on political prejudices as revolutionary thought gripped France at this time.

"The height of the mountains in Tahiti's interior is quite surprising and contrasts strongly with the relatively small size of the island. However, far from giving the island a sad and wild appearance these mountains serve to embellish the landscape by bringing under the eyes a different prospect with every step taken. It is a land everywhere covered with the rich lushness of nature's bounty; everything luxuriating in such magnificent disorder that the art of man could never imitate. From these lofty peaks descend a countless number of small streams; not only do they irrigate and fertilise the ground but serve also all the island's practical requirements while at the same time enhancing the beauty its landscape. All the level ground, from the seashore right up to the mountains, is given over to the cultivation of fruit trees, under these, as I have previously remarked, the Tahitians scatter their houses in a fashion that seems to be without plan or design; certainly not in a manner that could be considered a village. One could believe oneself to be at once amongst the Elysian Fields. There are footpaths laid out with practical intelligence which are maintained in the most fastidious manner to provide pleasant and easy communications between all parts of the region. The principal items produced on the island are: coconuts, bananas, breadfruit, yams, paw paws, okra, several other tuberous vegetables, fruits indigenous to the locality, a great deal of sugarcane (which is harvested where it grows in the wild state), a type of wild indigo and plant producing a very beautiful red and yellow dye, the name of which I do not know."

After arriving at Papeete and leaving my luggage at the Mahana Lodge hostel I took an excursion to valley with a relatively easily reached *Partula* population. For this I met up with Trevor Coote, the biologist who has been monitoring *Partula* populations for the past 14 years. We took the bus to Tiapa valley, this rather dilapidated vehicle provided views of the mountains, and most strikingly of the deep valley of Punaruu.

In the 18th century there were no real roads, just footpaths along part of the coast and running inland. In 1769 Joseph Banks recorded in his journal "Mr Green and myself went today a little way upon the hills in order to see how the roads were... We found as far as we went, possibly 3 miles, exceeding good paths and at the farthest part of our walk boys bringing wood from the mountains, which we look upon to be a sure proof that journey will be easy whenever we atempt to go higher."

Later the permanent coastal road was constructed. This 'Broom Road' was built by King Pomare II in the early 1800s. The name reputedly comes from the daily sweeping of the road by convicts. Herman Melville gave a description or it from 1842: "As we proceeded, I was more and more struck by the picturesqueness of the wide, shaded road. In several places, durable bridges of wood were thrown over large water-courses; others were spanned by a single arch of stone. In any part of the road, three horsemen might

BROOM ROAD, TAHITI.

The Broom Road in 1839 (United States Exploring Expedition)

have ridden abreast. This beautiful avenue by far the best thing which civilization has done for the island is called by foreigners "the Broom Road," though for what reason I do not know. Originally planned for the convenience of the missionaries journeying from one station to another, it almost completely encompasses the larger peninsula; skirting for a distance of at least sixty miles along the low, fertile lands bordering the sea. But on the side next Taiarboo, or the lesser peninsula, it sweeps through a narrow, secluded valley, and thus crosses the island in that direction. The uninhabited interior, being almost impenetrable from the densely wooded glens, frightful precipices, and sharp mountain ridges absolutely inaccessible, is but little known, even to the natives themselves; and so, instead of striking directly across from one village to another, they follow the Broom Road round and round."

Today the view remains much the same, but the road itself cannot be considered a "beautiful avenue". It is now a busy main road fringed by an almost unbroken line of buildings; houses, shops and warehouses. In places these extend inland a considerable distance, in others the mountains come almost to the road edge and the developed fringe is very narrow. At Tiapa valley we passed through one of the wider areas of housing, walking from the main road between modern houses towards the foot of the mountains. As the road narrowed to a track it entered the narrow valley. Here houses spaced out and gave way to patches of agricultural land. The valley sides were almost vertical, mostly with ferns clinging to the sides. The vegetated valley floor was a narrow strip of land either side of the small river. Walking further inland the agricultural areas give way to woodland. This was invaded by different species, but some areas retained natural character, with large Tahitian chestnut trees *Inocarpus fragifer*, known locally as 'mape'.

Walking up the valley we crossed and recrossed a fast flowing little river, fringed with more 'mape' trees. The riverbed is swept clear of debris by frequent floods, so this and most other Polynesian rivers are a jumble of smoothly rounded boulders. With little soil around them the rivers mostly run clear. The 'mape' trees fringe the rivers, with tall, almost straight trunks. There are few branches on the stems but high overhead a dense, dark canopy forms. On the stony ground the trees support themselves with large, sinuous buttress roots.

Leaving the suburban part of Tiapa valley

'Mape' in fringing the Tiapa river

After an hour we stopped at a patch of trees which Trevor knew was the home to *Partula incrassa*. This was a nondescript area of avocado, African tulip trees and bananas that is cleared periodically to plant new crops. In these areas agriculture is a very temporary thing; trees are cleared and bananas, manioc or other crops are planted. They are tended for a while and then abandoned. This cycle of clearance, regeneration and clearance again has probably been happening since the Polynesians first settled the islands 2,400 years ago, and contributes to the diverse but degraded nature of the lowland habitats.

At first glance these sorts of areas are not very interesting biologically; the vegetation is a jungle of introduced species with the exception of some of the ferns. There is generally little animal life other than some invasive ants and a small number of snail species that follow agriculture around the tropics. The conspicuously noisy birds were all the mynahs and bulbuls that have been introduced by man throughout the Pacific islands and beyond. The only exceptions were some Pacific swallows feeding in clearings, flying down from their nests and roosting sites in the cliffs of the valley walls. Somewhere further up the valley some of the last remaining Tahitian monarchs survive. Even these though are surviving in degraded habitat.

After a while of searching Trevor spotted a *Partula* on the underside of an avocado leaf just above head height. The tastefully shaded yellow-brown shell was quite conspicuous when pointed out, but difficult to spot in the dark canopy. Trevor was using a torch which made an enormous difference; he found three in total. I found nothing, putting my failure down to the persistent disorientation of the flight.

Partula incrassa is an odd species, not just for its survival in the abandoned agricultural habitat. It is one of the few *Partula* to have lived on more than one island, being descended from a population that used to live on Moorea. Henry Crampton thought it was a subspecies of the adaptable Tahitian *Partula clara*. It is now know to be a hybrid between *clara* and colonising snails from Moorea. It is still something of a mystery how these snails moved between the islands, and also how they formed new populations. As I was to see on Moorea later when visiting the reintroduction sites, getting tree snails to establish viable populations is far from easy.

Tiapa valley and *Partula incrassa*

After a night in Papeete I met up with Trevor again to make an excursion up into the mountains. The most easily accessible high forest snail populations are on Mt. Marau, behind Papeete. This is a steep mountain ridge running up to the crater rim of the island's old volcano. Because of its steepness, the upper slopes have mostly been spared from development, although there are pockets of agriculture and a radar station for the airport was built on the ridge.

For this excursion we were accompanied by Eric Lenoble who visits the mountains frequently as a tour guide, and Cindy Bick, a PhD student from Michigan University, working on snails. Eric drove us up the mountain road. I had walked up here 25 years before but thought I had failed to find the right road, stopping when I got to the rubbish dump. The real route carries on past the rubbish dump, turning into a track, and continuing for a long way through lush, but heavily invaded vegetation. The further up we went the more common tree ferns became, many with impressively tall trunks. Above these the ridges were covered in bracken fern, with isolated, twisted trees poking through.

Looking fromPapeete to Mt. Aorai (left) and Mt.s Marau (right)

The road winds along the moderately sloping western side of the mountain ridge, occasionally breaking onto the ridge itself. When it does this breathtaking views are revealed as the ridge drops almost vertically down many hundreds of metres to the eastern valley floor. The precipitous slopes are almost all covered in vegetation, most of which seems to be bracken ferns.

Looking down into the Punaruu valley

At several place we stopped to explore narrow ravines below the road. The tops of these were choked with invasive plants, including barbed raspberries. Below this tangle the ravines were filled with a mixture of plants; some ravines with many invasives like the velvet tree *Miconia calvescens*, bananas and some tulip trees, but also with lots of tree ferns and birds nest ferns. Other ravines contained ferns and other native plants. In most the large yellow *Partula otaheitana* snails were obvious on the birds nest ferns. Several other rare snail species were also present in good numbers.

At mid-day the drifting cloud settled on us and a cold rain started. I went with Eric to look at rare plants in the bracken of one of the read-side peaks. The climb was rather cold and miserable in the wind and rain, pushing through spiky bracken. These bracken areas are largely devoid of interesting plant or animal life but they do contain isolated plant rarities, relicts of once more widespread species.

After examining the plants we continued the drive up to the end of the road. This road had been cut 20 years ago for the construction of a communications tower for Papeete and the airport

Trevor searching a ravine for snails

below. This structure stands out at a viewpoint above the coast. The very end of the road has another tower structure and here the clouds lifted partially, revealing a fantastic view of the Mt. Maru ridge and the Diadem, although Mt. Aorai remained hidden in cloud.

Unsurprisingly the inland mountains have always tempted explorers, although in the early days they had very little success. Illness prevented Wallis from exploring Tahiti himself and the expedition's account of the island came from the mate's report "We began to climb the mountain while our old man was still in sight, and he, perceiving that we made our way with difficulty through the weeds and brush-wood, which grew very thick, turned back, and said something to the natives in a firm loud tone; upon which twenty or thirty of the men went before us, and cleared us a very good path; they also refreshed us with water and fruit as we went along, and assisted us to climb the most difficult places, which we should otherwise have found altogether impracticable. We began to ascend this hill at the distance of about six miles from the place where we landed, and I reckoned the top of it to be near a mile above the river that runs through the valley below. When we arrived at the summit, we again sat down to rest and refresh ourselves. While we were climbing we flattered ourselves that from the top we should command the whole island, but we now saw mountains before us so much higher than our situation, that with respect to them we appeared to be in a valley; towards the ship indeed the view was enchanting: the sides of the hills were beautifully clothed with wood, villages were every where interspersed, and the vallies between them afforded a still richer prospect; the houses stood thicker, and the verdure was more luxuriant. We saw very few habitations above us, but discovered smoke in many places ascending from between the highest hills that were in sight, and therefore I conjecture that the most elevated parts of the country are by no means without inhabitants. As we ascended the mountain, we saw many springs gush from fissures on the side of it, and when we had reached the summit, we found many houses that we did not discover as we passed them. No part of these mountains is naked; the summits of the highest that we could see were crowned with wood, but of what kind I know not: those that were of the same height with that which we had climbed, were woody on the sides, but on the summit were rocky and covered with fern. Upon the flats that appeared below these, there grew a sedgy kind of grass and weeds: in general the soil here, as well as in the valley, seemed to be rich. We saw several bushes of sugar-cane, which was very large and very good, growing wild, without the least culture. I likewise found ginger and turmerick, and have brought samples of both, but could not procure seeds of any tree, most of them being in blossom. After traversing the top of this mountain to a good distance, I found a tree exactly like a fern, except that it was 14 or 15 feet high. This tree I cut down, and found the inside of it also like a fern: I would have brought a piece of it with me, but found it too cumbersome, and I knew not what difficulties we might meet with before we got back to the ship, which we judged to be now at a great distance."

Cook's expedition was the first British naval vessel to include an official naturalist. Joseph Banks used his own considerable riches to equip a serious research team, including the great botanist and taxonomist Daniel Solander. Solander was probably responsible for collecting most of the animals and plants brought back by the *Endeavour* but unfortunately he did not keep a journal, so where he collected and what he observed are completely unknown. Banks himself was not very adventurous and his journal records few inland excursions. On 22nd May 1769 "This morning showery and cool, seemingly a good opportunity of going upon the

As Wallis found reaching the top of a mountain is a challenge

hills. I went accompanied only by Indians, indeed all of them but one soon left me, he however accompanied me during my whole walk. The paths were very open and clear till I came to the woods but afterwards very bad, so much so that I could not reach the top of the lowest of the two high hills seen from the fort, which was all I intended. I was in some measure however recompens'd by finding several plants which I had not before seen, with which I returnd before sunset..." Two days later "Msrs Monkhouse and Green atempted this day to climb the same hill that I attempted on the 22nd, with much the same success; they got however higher than I did but could not reach the summit."

Several weeks later Banks did undertake an exploration of the valleys inland. On 3rd July 1769 he recorded: "At length we arrivd at a place where the river was bankd on each side with steep rocks, and a caskade which fell from them made a pool so deep that the Indians said we could not go beyond it, they never did, their business lay upon the rocks on each side on the plains above which grew plenty of *Vae* [wild bananas]. The avenues to these were truly dreadfull, the rocks were nearly perpendicular, one near 100 feet in hight, the face of it constantly wet and slippery with the water of numberless springs; directly up the face even of this was a road, or rather a succession of long peices of the bark of *Hibiscus tiliaceus* which servd them as a rope to take hold of and scramble up from ledge to ledge, tho upon those very ledges none but a goat or an Indian could have stood. One of these ropes was near 30 feet in lengh. Our guides offerd to help us up this pass but rather recomended one lower down a few hundred yards which was much less dangerous, tho we did not chuse to venture, as the sight which was to reward our hazard was nothing but a grove of *Vae* trees which we had often seen before."

The US Exploring Expedition was much more intrepid "Dr. Pickering and Mr Couthouy being desirous of making another attempt to reach the top of Orohena... determined on attempting the ascent of the ridge leading directly up from Matavai Bay, as the one that had appeared to them most practicable... when they had reached the altitude of fifteen hundred feet they no longer found any paths; on arriving at this point, they halted for some time to make collections of land-shells, and some very interesting specimens were obtained of Helices, Partula, Cyclostomas, Curocollas, and Pupas; after this they continued ascending, the ridge gradually becoming narrower, until they reached as spot on the ridge where there was not room for one person to pass by another, and where they could look down a precipice on each side to depths of two thousand feet.

"Plants that were below of small size here grew into large woody shrubs; among them a species of Epacris was found growing luxuriantly along the crest of the ridges, and magnificent arborescent ferns on the mountains sides, some of them forty feet in height; another species was seen whose fronds were more than twenty feet in length. Their path was much impeded by the tangled ferns and wiry grass (Gleichenia), which it was impossible to get through without the aid of a knife or a hatchet. They had now reached four thousand five hundred feet..."

"Mr. Dana and some of the others, obtained leave of absence from Captain Hudson for five days with the design of ascending Mount Aorai. They commenced the ascent immediately in the rear of Papieti, and by noon on the second day had reached an elevation of five thousand feet, where they stood upon a platform

A grove of tree ferns

about twelve feet square... they pursued their ascending route along the edge of a ridge not more than two or three feet in width, having on each side an abyss two thousand feet deep. Seen from this ridge, looking south, Mount Aorai seemed a conical peak, but as it was approached it proved to be a mountain wall, whose edge was turned towards them. The only ascent was by a similar narrow path between precipices, and surpassed in steepness those they had already passed. The width of the crest seldom exceeded two feet, and in some cases they sat upon it as if on horseback, or were compelled to creep along it upon their hand and knees, clinging to the bushes. At last they reached the summit, where they found barely room to turn around."

Mount Aorai is indeed a stunning sight. The road on Mount Marau ends at a precipice facing Mount Aorai. Across the intervening valleys the cliffs of Aorai rise up much higher than the ridge of Mount Marau. It is rare for there not to be cloud on these mountains and so the view is fragmentary and fleeting. The steepness of the ridges produces odd effects on air circulation with winds being driven up from the valleys and meeting higher level

The ridge along Mt. Marau

winds blowing the other way. This produced an odd effect where cloud was rolling in from the east but being blown back by westerly valley winds, creating a clear cloud line.

Mt. Aorai to the left, hidden by cloud

Driving back down the mountain we stopped briefly to admire a few green fruit doves that landed in a tree in front of the vehicle. The first glimpse was a flash of dark green as one bird flew down the road, then when settled showing itself to be a pale creamy lime green to off-white on the head and breast and dark green on the back. These birds were among the first animals to be recorded in the islands; Wallis recorded "We saw great numbers of parrots and parroquets, and several other birds which were altogether unknown to us; we saw also a beautiful kind of dove, so tame that some of them frequently came close to us, and even followed us into the Indian huts." Bougainivlle noted "We saw there no quadrupeds other than pigs, a small but quite attractive species of dog and rats without number. The local people have domestic poultry which are in every aspect the same as our own. We also saw pretty greenish coloured turtle doves, large pigeons with beautiful royal blue plumage (these incidentally were exceptionally good eating) and very small parakeets or budgerigars with a plumage of the most striking mixture of red and blue. The pigs and poultry here are fed solely on bananas. "

Green fruit doves *Ptilinopus purpuraeus* (from Cassin 1858)

The parrots were seen on Cook's first voyage, when Sydney Parkinson recorded observations on the species that the expedition's naturalists collected:

"During our stay here, Mr. Banks and Dr. Solander were very assiduous in collecting whatever they thought might contribute to the advancement of Natural History; and, by their directions, I made drawings of a great many curious trees, and other plants; fish, birds, and of such natural bodies as could not be conveniently preserved entire, to be brought home.

"E ratta, or e pooratta. *Metrosideros spectabilis* [= *M. collina*]. This tree, or shrub, grows upon the Tooaroa, or Lower-hills, and is much resorted to by the venee, or small blue parrot, which feeds upon the flowers, and is often caught here, by means of a glewy juice which issues out from the tops of the stalks, when broke by their feeding upon them, and catches them like bird-lime: the flowers are full of beautiful scarlet stamina; the natives stick them in their ears by way of ornament; and the leaves are put in their monoe, when they can get nothing sweeter."

Although the parrots are long gone Meterosideros collina still survives on the hills, and the pigeons remain in reasonable numbers in the higher forest areas. Back in Papeete there are no pigeons other than the feral species that are found in towns all around the world.

With its concrete and air-conditioning there is relatively little wildlife. Even the mosquitoes that plagued my nights in Papeete in 1992 were absent. The islands have a reputation for being infested with mosquitoes. I did not find them particularly bad, but apparently they can be an infuriating nuisance at times. Historically flies were the greatest nuisance; in 1767 Wallis wrote "We saw no venomous creature here; but the flies were an intolerable torment, they covered us from head to foot, and filled not only the boat, but the ships." Banks in 1769 said: "The flies have been so troublesome ever since we have been ashore that we can scarce get any business done for them; they eat the painters colours off the paper as fast as they can be laid on, and if a fish is to be drawn there is more trouble in keeping them off it than in the drawing itself.

Meterosideros collina by SydneyParkinson

"Many expedients have been thought of, none succeed better than a mosquito net which covers table chair painter and drawings, but even that is not sufficent, a fly trap was nesscessary to set within this to atract the vermin from eating the colours. For that purpose yesterday tarr and molasses was mixt together but did not succeed."

Mosquitoes were more abundant on the next island, Maupiti, which has not yet been invaded by air-conditioning.

Maupiti

As is common in using Air Tahiti, my flight to Maupiti was delayed by an hour, which was plenty of time to get tired of the airport, however pleasant the domestic terminal is. The flight to Maupiti took 50 minutes with little to see due to thick cloud most of the way. As we flew westwards it thinned and broke, giving glimpses of Huahine, Tahaa and Bora Bora islands. Maupiti itself was cloudless, making for a strikingly beautiful first impression.

Contrasting cloudy Tahiti with arrival at Maupiti

Maupiti (along with Bora Bora) was the first of the Society islands to be seen by European explorers on 6th June 1722 when it was sighted by Jacob Roggeveen. The first European landing was over 100 years later, on 9th June 1823 with the arrival of *La Coquille.* Neither expedition (nor any subsequent ones) made any comment on the island, just recording its presence and their visit.

Maupiti's airstrip is on one of the motus, the small fringing coral islands. The motu is covered in typical Indo-Pacific beach crest plants behind a narrow beach, beyond which the ocean waves break almost immediately. Sea-birds could be seen over the lagoon; a frigatebird and a number of crested terns. On the island there were barred ground doves and chickens but, surprisingly, no mynahs.

The little airport comprises one building, just a shelter from the sun and rain. On the far side of this the boats wait for the passengers. On Maupiti I was met by the owner of my 'pension' who took me on a drive around the island. Being a small place this only took a few minutes. The touristic highlight seemed to be the island's only sandy beach; a sand spit of a few metres. The rest of the coast is fringed by mud. Typical tropical island beach tourism requires a trip to the motus.

Boats waiting at the airport motu

Maupiti is the island least affected by development. Although it has all the modern infrastructure of a tarred coastal road, cars, truck and motorbikes, modern housing and good internet access, it has no hotels. All the visitor accommodation is in the local 'pensions', rooms in private houses. Mine was typical; a bit rough although with internet. The pensions provide excellent local food, with plenty of varied fish and breadfruit.

All the roadside gardens have their own breadfruit trees. These luxuriant, well tended trees come in several varieties differing most obviously in having divided or entire leaves, but also varying in some two dozen fruit characteristics. Almost everywhere else in the world there is just the one variety of breadfruit, which can be attributed to Captain Bligh.

For his second voyage Captain Cook selected William Bligh as Master. Under Cook's guidance Bligh became one of the most able captains of his time, and certainly the most remarkable navigator the British navy has ever produced. In 1788 he returned to Tahiti as captain of his own vessel, the infamous HMS *Bounty*. His orders were to transport seedlings of the invaluable Polynesian breadfruit tree to the Caribbean where it was to form a food supply for the plantation slaves. In October 1788 they landed on Tahiti and

Maupiti in 1826, by Duperrey 1838

remained there until 4th April 1789. This long residence was forced on them by the necessity of establishing the seedlings in pots in the captain's cabin before sailing. It also led the crew to form bonds with the people on the island, particularly the women.

The dangers the Polynesian women posed to naval discipline was apparent from the very first expeditions. Bougainville's arrival in 1768 was greeted with canoes "full of women whose faces were as beautiful as those of any European women; their magnificently formed and proportioned bodies would rival advantageously any of their sex no matter who or where in the world. The great majority of these beautiful creatures were completely naked because the men and the older people accompanying them had removed the mantles they usually wore. From the very first, but, only from the sanctuary of their canoes, they made in all innocence coquettish gestures and expressions towards us. Despite their apparent naivete however, we thought we could sense that they felt in some degree timid and embarrassed at

Breadfruit (*Artocarpus altilis*) by Parkinson in 1769

being so flagrantly exhibited in the presence of perfect strangers. ...

"The men openly encouraged us to choose a girl and to follow her onto the island; their gestures, the meaning of which it was impossible to misinterpret, only too clearly demonstrated the manner in which they expected us to make their acquaintance. I posed myself the question; how was I to restrain at their places of duty, in the middle of a spectacle such as this, 400 young French sailors who during the preceding 6 months had not even had sight of a woman?

"Thus it was, despite all our best efforts and the precautions we took, that one of these young girls managed to scramble aboard and make her way to the quarterdeck where she sat herself down on the hatch coaming just forward of the capstan. This hatchway at that time was open in order to provide free circulation of air to the men working at the capstan bars below. The young girl negligently allowed the mantle covering her to slip from her shoulders and revealed to the eyes of the entire crew the glorious beauty of her young body. This provocative gesture was done in exactly the same manner as when Venus had shown herself to the Phrygian shepherd. This parallel is indeed apposite for every aspect of this maiden was truly celestial. Sailors and soldiers fell over themselves in their haste to encircle the hatchway; never in the annals of all maritime history has any capstan ever been turned with such enthusiasm!

"Eventually our efforts to get the sailors under some sort of control did meet with limited success. However, the greatest difficulty I was confronted with at a personal level, and I must freely admit it, was in subduing my own over-stimulated amorous inclinations!"

Banks was apparently even more disturbed by the unexpected customs of the islanders: "One amusement more I must mention tho I confess I hardly dare touch upon it as it is founded upon a custom so devilish, inhuman, and contrary to the first principles of human nature that tho the natives have repeatedly told it to me, far from concealing it rather looking upon it as a branch of freedom upon which they valued themselves, I can hardly bring myself to beleive it much less expect that any body Else shall. It is this that more than half of the better sort of the inhabitants of the Island have like Comus in Milton enterd into a resolution of enjoying free liberty in love without a possibility of being troubled or disturbd by its consequences; these mix together with the utmost freedom seldom cohabiting together more than one or two days by which means they have fewer children than they would otherwise have, but those who are so unfortunate as to be thus begot are smotherd at the moment of their birth. Some of these people have been pointed out to me by name and on being askd have not denyd the fact, who have contracted intimacies and livd together for years and even now continue to do so, in the course of which 2, 3 or more children have been born and destroyd."

While all the expeditions experienced lapses of discipline, it was on the *Bounty* that this reached critical levels. The mutiny that followed resulted in Bligh and the crew loyal to him being cast adrift and the breadfruit seedling being thrown overboard. Whether he was a good manager of people or not, Bligh was remarkably determined and after reaching land he returned to Polynesia to finish his mission. Of the breadfruit he collected on his second expedition, one variety proved very successful and now grows on islands throughout the tropics.

The other plant that is unusually diverse in Polynesia is the banana. This also has a long history in the islands; Sydney Parkinson

Sydney Parkinson's sketches of the Tahitians in 1769

described the banana in 1869 as "Meiya. *Musa-paradisaica*. - This is the well-known tropical fruit called Plantains, and Bananas, of which there is a great variety in these islands: they reckon more than twenty sorts which differ in shape and taste; some of these are for eating raw, and others best boiled, and will serve instead of bread: they plant them in a rich soil, and take great pains in their cultivation." Banks gave the number of varieties as 13 "the best I have ever eat" as well as plantains "but indiffer[e]nt".

Night-time on tropical islands can often be noisy as every house has several dogs and cockerels, all of which see the night as the time to declare their territories. On Maupiti the dogs and cockerels were joined by the low barking of *Hemidacylus* geckos. These can also be heard on Tahiti but are not so conspicuous in the modern, air-conditioned buildings of Papeete but are present in the road-side trees.

The day starts early on Maupiti, with the dawn chorus from the cockerels and the working day starting soon after. This day I took a departure from snail hunting to join the pension's excursion along with the other guests, to swim with manta rays and visit one of the motus. We took the pension's boat out to one of the 'manta stations' where they visit every morning for the attentions of the cleaner wrasses. There were a few boats there and many people in the water. By the time we had got into the water the mantas had moved on, so began a morning of following the rays from station to station, usually slightly too late. Finally we found one cruising around in water about four metres deep. It was moving very languidly and drifting around in a relatively small area until someone swam down to it when it slowly drifted away from us. Clearly they don't like being hassled, but don't show it.

As well as swimming with the mantas we visited a so-called 'coral garden', a shallow area with coral rocks and outcrops for snorkelling with the fish. In the past few decades the reefs have been badly damaged by coral bleaching but here there was still a fairly good percentage of live coral. Fish were abundant and diverse, with many damselfish and coralfish in particular.

From there we went through the pass into the open ocean (which was quite calm) to look for whales near the pass, unsuccessfully.

Humpback whales *Megaptera novaeangliae* migrate to these islands every year to breed. Being volcanic, the islands rise from the seabed extremely steely, so deep water is very close to the shore. This means that whales come remarkably close to the reef, making it an exceptional place to watch them.

Human interest in whales in the area did not used to be so benign. Once the islands had been mapped, and navigation in the region became reliable, the whalers moved in. From 1775 small numbers of whales were being caught but whaling could not be considered an industry until 1788 when the British ship *Amelia* entered the Pacific. Between 1802 and 1846 730 American whalers were operating in the south Pacific, taking sperm whales, southern right whales as well as humpbacks. The whaling boom came to an end in the 1860s when economic collapse and civil war put an end to the American operations and the development of petroleum oil refining in 1859 cut the demand for whale oil.

A manta ray

The Society islands were relatively minor players in the whaling industry, being home mainly to humpbacks which were in less demand than sperm whales. However, the islands were a stopping-off place for the whalers. Among the crews of these visiting ships were two notable names, Andrew Garrett and Herman Melville.

Garrett had gone to sea at the age of 16 as a deckhand. In 1846 he was in the Pacific, on the *Eliza L.B. Jenney*, reaching Hawaii. He settled in the islands in 1852 and became a notable shell collector. He spent next 35 years collecting shells for various museums and private collectors. He explored almost all island groups of the Pacific, collecting and painting shells, sea-slugs and fish.

This work led to him visiting the Society islands in 1857. On this voyage he was a passenger on the whaler *Lydia*. The ship's captain John W. Leonard wrote that they had passed the "Island of Huehine [sic] where we have gathered a splendid lot of shells and fishes. We have a naturalist as Passenger with us which makes it very pleasant. And he is painting a fine lot of fish for me and collecting a beautiful lot of shells. So I shall have a fine Colection [sic] when I get Home." The collection was transferred to the *John*

A curious damselfish

Gilpin which tragically was wrecked and the bulk of the collection lost. Garrett gave a description of his approach to collecting on the *Lydia*: "When searching along the coast I have to take pencil and paper, an assortment of small jars, boxes and calabashes which my native boys carry. And when I find anything I wish to preserve I first note the depth of water, kind of bottom, its mode of locomotion and colors while alive. And I find it necessary to preserve them in water while carrying them about so that their delicate parts will remain perfect until I can place them in alcohol."

Garrett returned to the Society islands in 1860 for a three year exploration of the islands. Although he was not impressed by the islanders' lives where "indolence, drunkeness, and the most loathsome diseases" were rampant, he settled on Huahine, living there from 1870 until his death in 1887. He had been taken by the island on his first acquaintance with it back in 1858 on the *Lydia*, writing that the two parts of the island "present a bold and mountainous aspect, and are clothed in the most luxurious verdure from the water's edge to the summits.... A short distance back there arises and amphitheatre of hills and mountains which are covered either with tall, rank grass or dense dark forests, and, the whole coast consists of a dense mass of fruit and splendid flowering trees, all combining to form one of the

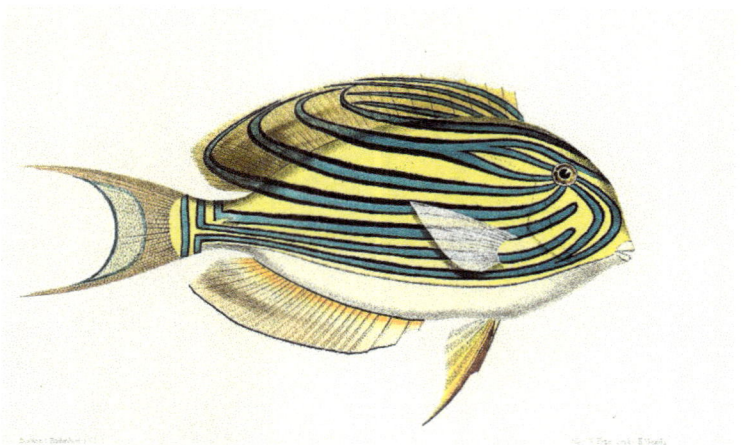

One of Garrett's fish from his Fishe der Sudsee (1873)

most delightful tropical scenes I ever witnessed."

Herman Melville's whaling career was even briefer, signing on to the *Acushnet* in 1841, but jumping ship in the Marquesas Islands in 1842. His experiences in the Marquesas led to his first book, 'Typee', in 1846, followed a year later by 'Omoo', his account of sailing from the Marquesas to the Society islands. In a further two years in the Pacific, Melville spent another six months aboard a whaler. These two brief periods aboard the *Acushnet* and the *Charles & Henry* were used as the basis for 'Moby-Dick'.

We stopped at the motu to the east of the pass where all the boats had gathered for a communal Tahitian 'barbecue'. The islanders had used the traditional cooking technique of burying breadfruit, plantain, chicken, pork and clams on top of hot rocks earlier in the morning. The meat was good but the vegetables very stodgy and dull. There was also a sweet concoction made of banana stewed in coconut that was sticky and quite unpleasant. I came across this later, and found that it has a good flavour if cooked carefully, although it is still a disconcerting, very sticky yellow snot.

Food prepared this way seems to be something of an acquired taste. Back in 1768 Joseph Banks commented "I speak now only of what is more properly calld Fish; but almost every thing which comes out of the sea is eat and esteemd by these people,Shellfish, lobsters,

The motus

Crabbs, even Sea insects and what the seamen call blubbers [jellyfish] of many kinds conduce to their support. Some of the last indeed that are of a tough nature are prepard by suffering them to stink; custom will make almost any meat palatable and the women especialy are very fond of this, tho after they had eat it I confess I was not extreemly fond of their company. ...

"Cookery seems to have been little studied here: they have only two methods of applying fire, broiling, or baking as we calld it which is done thus. A hole is dug in depth and size according to what is to be prepard seldom exceeding a foot in depth, in this a heap is made of wood and stones alternately laid; fire is then put to it which by the time it has consumd the wood has heated the stones sufficiently just enough to discolour any thing which touches them. The heap is then divided; half is left in the hole the bottom of which is pavd with them, on them any kind of provisions are laid always neatly wrappd up in leaves, the whole is then coverd with leaves on which are laid the remaining hot stones then leaves again 3 or 4 inches thick and over them any ashes rubbish or dirt that lays at hand. In this situation it remains about 2 hours in which time I have seen a midling hog very well done, Indeed I am of opinion that victuals dressd this way are more juicy if not more Equally done than by any of our European methods, large fish more especialy. Bread fruit cookd in this manner becomes soft and something like a boild potatoe, tho not quite so farinaceous as a good one yet more so than the midling sort. Of this 2 or 3 dishes are made by beating it with a stone pestil till it make a paste, mixing water or Cocoa nut liquor with it and adding ripe plantains, bananas, sour paste &c.

"As I have mentiond Sour paste I will proceed to de[s]cribe what it is. Bread fruit by what I can find remains in season only 9 or 10 of their 13 months so that a reserve of food must be made for those months when they are without it. To do this the fruit is gatherd when just upon the point of ripening and laid in heaps where it undergoes a fermentation and becomes disagreably sweet; the core is then taken out which is easily done as a small pull at the stalk draws it out intire, and the rest of the fruit thrown into a hole dug for that purpose generaly in their houses; the sides and bottom of which are neatly lined with grass; the whole is coverd with leaves and heavy stones laid upon them. Here it undergoes a second fermentation and becomes sourish in which condition it will keep as they told me many months. Custom has I suppose made this agreable to their palates tho we dislikd it extreemly, we seldom saw them make a meal without some [of] it in some shape or other."

Fortunately I never came across this "Sour paste".

The edge of a motu

The distinctive Polynesian method of mooring boats

Being woken by cockerels at dawn helps maximise time for field work, but even so I was not the only visitor on the mountain path the next morning. Some were even descending already. The way up was steep, very dry and hot. The vegetation was very poor: mango trees,

Back on Maupiti

sparse shrubs and undergrowth of ferns – all garden escapes. This was not encouraging habitat for finding snails and the first interesting find was a predatory flatworm. It was only near the top that I found snails. Insect life was also scarce; with one large moth, very few ants and several introduced wasps being seen.

The vegetation near the top was much more interesting, with *Pandanus* screw pines and *Hibiscus* trees, although still with scattered mango trees. The *Pandanus* were full of minute snails in an area that was distinctly cooler than elsewhere on the hike.

The ridge itself was the obvious attraction for the hikers. Rather than join other people on the small, precipitous viewpoint I continued upwards as a there was second peak only a little further on. The view from there was spectacular, hidden to one side by the vegetation of the continuing ridge, but clear on all other sides. Bora Bora could just be seen in the haze on the horizon.

There was a path continuing up the ridge. This was clear but also clearly not used by the tourists (in fact I didn't see anyone go beyond the first viewpoint). The path soon reached a third peak,

Pandanus thickets

and continued further. This stretch was obviously very infrequently used. Some maps indicate a cross-island route, so I continued. The vegetation had deteriorated again with dense mango leaves covering the path. The way was far from clear but could be followed by an impression of the route and some broken plant stems where someone had passed in the last few days. It started to descend more steeply but glimpse through the trees suggested a disturbingly steep descent. This was confirmed by my arrival just above a cliff, marked by a cloth fluttering from a branch. If there was a way down from there it was going to be extremely difficult, so returning was the sensible course.

Back on the second peak I watched tropicbirds flying among the cliffs and the large numbers of orange dragonflies flying around me. As I had thought mynahs and bulbuls were absent although I could hear white-eyes. The only birds I saw other than the tropicbirds were a few very nonchalant chickens on the path and down at the coast barred ground doves and feral pigeons (and lots more chickens).

On the descent there were many blue-tailed skins running about. Later, on other islands, I was to find that they were almost

View from the top of Maupiti

always to be found in the afternoon, when there were fewer people around and when the sun had moved from its zenith, with more shade cast on the ground.

On these small, isolated islands weather patterns move in rapidly and with little warning. The mountain hike day was ideally clear and dry but the following morning started with monsoon rain. The rain lessened after dawn but drizzle remained throughout the day. I did not appreciate the significance of this until I arrived back at the airport motu. There the speculation started: would the aeroplane be operating or not? Maupiti has a short runway and it seems that pilots are sometimes unwilling to land there when it is wet. Several people said that it's fine if the pilots are Tahitian but the French ones are more cautious. Frustratingly it seems that our pilot was French. As scheduled the plane did stop at Bora Bora, just visible

Blue-tailed skink

through the drizzle, and waited there for the weather to lift. We waited for a decision, and just as the flight was cancelled the sun came out.

The following day was grey but not actually raining, and forecast to improve. Back on the airport motu the now familiar group of passengers waited. The newly scheduled flight was delayed, but that is not unexpected with Air Tahiti. However, once past the scheduled departure the weather started to worsen. There followed an anxious wait of weather watching. Fortunately either the weather didn't seem too bad from the air, or this pilot was Tahitian, for we did managed to leave, only just over 24 hours late.

The flight gave no views other than a glimpse of Raiatea and eventually Moorea looming out of black cloud and a few minutes later the fringe of Tahiti, similarly dark.

A wet day on Maupiti

Moorea

The extraordinary appearance of Moorea was noted by the United States Exploring Expedition: "Eimeo [Moorea was variously called Eimeo or Imao at the time] has, if possible, a more broken surface than Tahiti, and is more thrown up into separate peaks; its scenery is wild even in comparison with that of Tahiti, and particularly upon the shores, where the mountains rise precipitously from the water, to the height of twenty-five hundred feet."

My visit to Moorea was more rushed than intended, having lost a day on Maupiti. I met up with Trevor and Cindy who were able to provide transport to the interesting *Partula* locations. We took the car ferry from Papeete harbour to Vai'are bay. The previous couple of day's wet weather had moved on but on the outside deck of the ferry was extraordinarily windy. Despite the wind it was worth standing there to watch crested terns flying alongside the ship and diving vertically to take fleeing from the ship's wake.

The road hugs the coast and the most direct route to the snails was around the north-eastern point of the island. We drove round the north-east side of the island, stopping to drop off my

luggage at the Painapaopao Backpacker hostel. The northern slopes are very dry and clearly degraded habitat, in contrast to the lush inland forests.

On the north coast two great bays dissect the island. The eastern one is Cook's Bay. This is named after Captain Cook, marking his visit to the island in 1777 in the *Resolution* and *Discovery*. He actually anchored in the western bay, today called Opunohu bay. Although his first expedition visited the island in 1769, Cook himself remained on Tahiti. Banks was with the party that landed on 3rd June 1769 when he "repaird to the Island, where I could do the double service of examining the natural produce and buying provisions for my companions who were engagd in so usefull a work. .... I spent the rest of the day in examining the produce &c. of the Island and found it very nearly similar to that of Otahite, the people exactly the same, indeed we saw many of the Identical same people as we had often seen at Otahite, and every one knew well every kind of trade we had and the value it bore in that Island. The hills in general came nearer to the water and flats were consequently less, and less Fertile, than at Otahite - the low point near which we lay was composd intirely of sand and coral.

"Here neither Breadfruit nor any usefull vegetables would grow; it was coverd over with *Pandanus tectorius* and with these grew several plants we had not seen at Otahite, among them Iberis , which Mr Gore tells me is the plant calld by the voyagers scurvy grass which grows plentifully upon all the low Islands." The place where Banks landed was Afareaitu, just south of the ferry port of Vaiare.

Cook's anchorage at Opunohu Bay was our first stop. Here a tiny patch of mangrove ferns, circumscribed by the road, an inlet and a shrimp farm, is occupied by a population of *Partula taeniata* snails. These were easy to find and remarkable in their relative abundance, colour variation and unusual habitat (living in the mangroves rather than in forest).

Banks explored the coasts of Moorea quite extensively but recorded little of interest, as he noted himself "After having slept last night without the least interuption we proceeded forwards but during the whole day saw little or nothing worth observation" (30th June 1769).

HMS *Resolution* and *Discovery* in Opunohu Bay by John Cleveley, print by Thomas Martyn ,1787

Partula taeniata and the mangrove fern habitat

From there we drove up to the Belvedere area, through agricultural land. Just below the Belvedere are the old marae in forest habitat. The marae are the most conspicuous archaeological structures in Polynesia. They comprise flat or raised pavements of rock, sometimes stepped, sometimes surrounded by vertical sheets of rock. They were the site of formal and ritual occasions, ranging from decision making to religious events. Captain Cook witnessed a human sacrifice at one on Tahiti, but probably their use was usually more benign than that and their greatest significance would have been as a marker of land ownership.

The oldest ones date back 1,200 years, but others are only a few hundred years old. On Moorea the older ones are coastal, but as the human population grew (around 700 years ago) marae started appearing inland. In the Opunohu valley area 220 marae and associated structure have been identified, the most notable of which are just below the Belvedere. Here several areas of low, solidly built rock walls form platforms or enclosures. They would be a strange sight on their own, but with the splendidly buttressed mape trees growing in them in rows they are decidedly peculiar.

From the marae we walked through forest to various points where *Partula tohiveana* had been reintroduced in 2016. This forest of mape is relatively easy to move through, the trees being well spaced and mostly lacking branches for a considerable height. For a Polynesian site the ground is fairly flat and is covered in small pebbles and rocks, regularly washed into temporary heaps by the flood waters that rush through the river-bed down to Opunohu bay.

We did not find any *Partula* in the reintroduction sites until we reached the last one. This was the site of the very first attempts at reintroducing snails to the wild, back in 1995. There an extraordinary reserve was created, with an electric fence and salt filled trough to exclude the predatory *Euglandina* snails. Unfortunately the predators always managed to find a way through the defences and the reserve had to be abandoned. The reserve is a remarkable sight, with a very dense low growth of *Freycinettia* which restricts itself to the confines of the reserve, even though the fences have gone. Searching

Opunohu valley was agricultural land even in 1849 – E.G. Fanshawe, National Maritime Museum, Greenwich, London

Forest near the Belvedere

The old *Partula* reserve

in the reserve Cindy found a juvenile *Partula tohiveana*, being a juvenile this snail was not one of the ones released in 2016 but one of the first born on Moorea.

After continuing from the reserve to the Belvedere I left the others to return to Tahiti while I carried on up the mountain to the Trois Cocotiers pass. The lower parts of the path were fairly poor but there were areas of good mape forest further long. The ascent to the ridge was seep and the ridge itself exposed to the extremely strong wind. From here there were even more of Moorea's stunning views, in an almost complete 360 degree panorama.

I took a smaller path along the ridge, through bracken fern, past the young trees that replace the three coconuts that used to give 'Trois Cocotiers' its name until a hurricane blew them down. Near the foot of the next peak I found a shell of a juvenile *Partula taeniata* showing that some snails persisted in the forest.

Herman Melville has often been accused of embellishing

Moorean born *Partula tohiveana*

Mt. Rotui and Opunohu Bay (left) from the Belvedere

Cook's Bay from the Belvedere

the accounts of his stay in the Marquesas and Society islands, but his island descriptions ring true at least. Of Moorea he wrote: "We were in the valley of Martair; shut in, on both sides, by lofty hills. Here and there, were steep cliffs, gay with flowering shrubs, or hung with pendulous vines, swinging blossoms in the air [the profusion of flowers may well be a fiction]. Of considerable width at the sea, the vale contracts as it runs inland; terminating, at the distance of several miles, in a range of the most grotesque elevations, which seem embattled with turrets and towers, grown over with verdure, and waving with trees. The valley itself, is a wilderness of woodland; with links of streams flashing through, and narrow path ways, fairly tunneled through masses of foliage.

"At the foot of the mountain, a steep path went up among rocks and clefts, mantled with verdure. Here and there were green gulfs, down which it made one giddy to peep. At last we gained an overhanging, wooded shelf of land which crowned the heights; and along this, the path, well shaded, ran like a gallery. In every direction, the scenery was enchanting. There was a low, rustling breeze; and

View from Col des Trois Cocotiers (one coconut visible to the right)

below, in the vale, the leaves were quivering; the sea lay, blue and serene, in the distance; and inland the surface swelled up, ridge after ridge, and peak upon peak, all bathed in the Indian haze of the Tropics, and dreamy to look upon. Still valleys, leagues away, reposed in the deep shadows of the mountains; and here and there, water-falls lifted up their voices in the solitude. High above all, and central, the "Marling-spike" lifted its finger...

"Following another path, in descending into the valley, we passed through some nobly wooded land on the face of the mountain. One variety of tree particularly attracted my attention. The dark mossy stem, over seventy feet high, was perfectly branchless for many feet above the ground, when it shot out in broad boughs laden with lustrous leaves of the deepest green. And all round the lower part of the trunk, thin, slab-like buttresses of bark, perfectly smooth, and radiating from a common center, projected along the ground for at least two yards. From below, these natural props tapered upward until gradually blended with the trunk itself [this was clearly 'mape']."

After exploring the ridge I returned to the Belvedere, and

down through the agriculture and across the Route des Ananas to Cook's Bay. As the name suggests, this road passes through several pineapple fields, as well as mahogany plantations. The most interesting sight on this walk was a pair of harriers circling over the valley. Western marsh harriers *Circus aeruginosus* were introduced to Tahiti in about 1884 by the German Consul to eat the rats that swarmed over the islands. Polynesian rats *Rattus exulans* had arrived in the islands with the Polynesian seafarers many hundreds of years ago, black rats *R. rattus* came with the European explorers, sometimes in dramatic numbers. On Cook's second voyage the *Resolution* was "a good [deal] pestered with rats" he wrote. On Moorea "I hauled her within thirty yards of the Shore, being as the depth of water would allow." The midshipman "got a Hawser out of the Ballast Port with some Spars lash't upon it with a desire to get some of the Rats out of the Ship, we having a Great Number of them on board." The harriers did little; they do eat rats and mice, but also other birds. The harriers were not common until the 1970s and pairs of them circling in the sky are now a frequent sight on most of the islands.

Paopao's shopping street

At the village of Paopao on Cook's Bay I shopped for supper and met the owner of the hostel who drove me back there. The hostel was very simple but comfortable and quiet. There are good views out to sea, with its unbroken horizon. It is probably a good spot if you are interested in the sea, but the northern headlands have been burnt many times in the past and seem to have no indigenous animals. Similarly the only native plants are those of the strand-line.

The following morning I took an early walk from the hostel to the airport for my flight to the next island. My route retraced the road we had driven along the day before, through the main tourist areas of Moorea. The buildings were smarter than elsewhere on the island, although a few old beach-side shacks remained amongst the hotels, shops and car-hires. It was a pleasant enough walk although there is little of particular interest along this road, other than the views and the stretch of rocky shore is good for slaty egrets searching for fish in the shallows.

Moorea's airport is at the north-east corner of the island, on what was once a sandy coastal plan on top of the ancient reef-flat. Between this and the mountains lies a lagoon. This was described in Wallis' first sighting of the island: "I steered S.W. by W. close along the north east side of it, but could get no soundings: this side is about six or seven leagues long, and the whole makes much the same appearance as the other, having a large salt water lake in the middle of it."

Along the coastal road

Huahine

The flight from Moorea to Huahine takes just half an hour. There is nothing to look at on the way, other than featureless ocean. As the plane comes in to land at Huahine it approaches from the south-east, giving good views of both the south and north parts of the island, and the inlet between them. Huahine is two islands: Huahine-Nui and Huahine-Iti, joined by a narrow, stretch of low land.

From the very small but smart airport I walked into the island's capital town of Fare. Right by the airport is a *Pandanus* forest growing over part of the tidal lagoon. This is a very unusual habitat; in most places the plants grow singly but in swampy places they can sometimes form small forests. Their stilt roots support them above the soft ground.. In these environments they often support a range of strange specialist animals such as flattened bugs and beetles that squeeze between the leaves. I'm sure there was plenty of interest to find in the airport *Pandanus* forest but it is a horribly spiny plant, with vicious hooks to ward off the over-inquisitive biologist. I left its mysteries for someone else to investigate.

The lagoon between the airport and Huahine-Nui

Pandanus forest

I checked in at the guest house Chez Guynette by the harbour in Fare. From there I went straight out to look for Andrew Garrett's grave. The small historical graveyard was along the coast, just behind the tiny stretch of sand that passes for Fare's beach. Here a gravestone was visible in crudely wired area. This wasn't Garrett's grave but I found it among some others hidden in a dense clump of *Hibiscus* trees and partially overgrown with creepers.

That evening the horizon was a bit cloudy so there was no sunset behind the islands visible to the west: Tahaa, Raiatea and more distant Bora-Bora.

For my three days on Huahine my intention was to climb as far up the island's mountain as I could reach. A *Samoana* snail had been photographed on the summit of Mt. Turi some 10 years before and I wanted to see if they were still there.

Andrew Garrett's grave

Joseph Banks had attempted to explore the mountain but did not have much success, The *Endeavour* had anchored in "Owalla" bay on 16th July 1769 and on the following day he "Went ashore this morn and walkd up the hills; found the productions here almost exactly similar to those of Otahite; upon the hills the rocks and clay were burnt if any thing more than they were in that Island. The people also were almost exactly like our late [friends] but rather more stupid and lazy, in proof of which I need only say that we should have gone much higher up the hills than we did if we could have perswauded them to accompany us, whose only excuse was the fear of being killd by the fatigue."

He says nothing more about the inland excursion and restricts his description to the coast: "flats were filld with very fine breadfruit trees and an infinite number of Cocoa nuts, upon which latter the inhabitants seemd to depend much more than those of Otahite; we saw however large spaces occupied by lagoons and salt swamps upon which neither breadfruit nor Cocoa nuts would thrive."

Mt. Turi from Fare town

Banks found the island's nature to be much the same as Tahiti, recording on 19th July "In all our searches here we have not found above 10 or 12 new plants, a few insects indeed and a species of scorpions which we did not see at Otahite." The scorpion was presumably the cosmopolitan lesser brown scorpion *Isometrus maculatus* or the Pacific dwarf wood scorpion *Liocheles australasiae*, both of which are occasionally found and tend to cause a moment of unnecessary panic in the local press.

There is no marked path up the mountain so the start of the hike was very much guesswork. The suburban roads turn to tracks through agriculture. Further up the valley sides the agriculture makes way for secondary forest tangled with creepers and patches of planted bananas. The occupied areas seemed to cease but what was now a narrow footpath carried on. A branch to the left looked like it might cut up to the ridge I was aiming for but the right looked like a more convincing path. This crossed a river bed and continued up the slope. There was an area that wild pigs had churned up, some feral chickens pretending to be forest birds and then another banana plantation. This came to a fence, over which there was an indistinct track heading straight up towards the ridge through secondary forest of coconuts and mangoes.

A view of Huahine by John Webber in 1783 (National Maritime Museum, Greenwich, London)

The path became steeper and eventually disappeared in a stream bed running vertically down the slope. At this point there was a large mape tree growing on the rock of the steam bed, its roots creating an overhang. Lying under the overhang was a *Samoana* shell, making the survival of the species considerably easier than anticipated (having taken just an hour).

Although there were good numbers of mape trees at this level the way up got harder from then on, with increasing density of *Hibiscus* stems. Although lots of *Hibiscus* is generally taken as a good sign in *Partula* research as the snails used to favour this tree, it can be a great annoyance in field-work. A few trees are fine; they tend to be sprawling things, somewhere between an untidy tree and an enormous bush. You can't walk under them, you have to climb though them, but that is usually quite easy. A *Hibiscus* forest is another matter. Here they grow densely and you very quickly tire of the gymnastics needed to get between the stems. This particular forest seemed to be exceptionally difficult to get through.

After a while I realised it was so difficult because it was

The track up the mountain

now a mix of *Hibiscus* and the thick stemmed *Merremia* creeper. However, by this point I could see light through the vegetation and soon made it to the ridge. This narrow ridge was about two metres wide, with a clear path running along it. Presumably this had come up from the fork that I had rejected earlier. Vegetation on the ridge was considerably better than on the slopes, with a mixture of *Hibiscus*, mape and an increasing diversity of native trees, interspersed with areas of elephant grass.

Merremia is a characteristic plant of the lowlands of almost all Indo-Pacific islands. It seems to be regarded as an invasive species everywhere, but it has long had a very wide distribution; the 18th century naturalist explorers found it in on uninhabited islands in the western Indian Ocean, in south-east Asia and on Tahiti. Maybe Polynesian sea-farers moved it around or maybe it spread throughout the Indo-Pacific naturally. Either way, it is highly invasive because it thrives in the open, disturbed areas created by clearing land for agriculture or cutting roads through forest.

Hibiscus and *Merremia* tangle and Parkinson's illustration of *Merremia* collected on the *Endeavour*

The path came to a gigantic giant tree, an extraordinarily large mango tree. The open area around the enormous trunk provided a glimpse of the peak ahead, but from there it was difficult to tell if it was the peak of Mt. Turi, the false peak beside it or the top of a cliff. Further on the vegetation was wetter, with lots of ferns and moss, clubmosses and orchids. Invasive purple ground orchids were flowering in profusion, but endemic epiphytic species were also present. In this lower, patchier forest more views of the top could be seen.

I had escaped from the *Merremia* some time ago and there was also little *Hibiscus*, both really being lowland plants. They were in the valleys on either side of me but not on the ridge. There were a few trees along here, but much of the tall vegetation was made up of ferns and *Freycinettia*. Walking was quite easy, although slippery in places. Spiders were more common here than I'd noticed elsewhere. Almost all of them a species of small kite spider pretending to be a piece of bark hanging in a web. Later I found that far from being an interesting part of the native fauna, as I'd assumed, this is another invasive species. The spiny-backed spider *Thelacantha brevispina* has been found from Madagascar, though Australasia to Polynesia. It seems to have reached the Society islands in the middle of the last century; there is an old unsubstantiated record, but the earliest definite records are from 1959, also from Huahine. Strangely it seems to be well established on Huahine but scarce elsewhere in these islands.

At mid-day I reached the top – an area of bracken with some shrubby trees and the ends, and a radio mast with solar panels in the middle. To one side was the slight higher peak of Mt. Turi.

Between us was a drop down to the saddle, which looked like a possible but unnecessarily hard climb.

While I rested there I watched a harrier circling overhead. I could also hear fruit doves in the trees below, but other than those there were no birds; chickens do not go up to the ridge.

Descending the ridge was faster, but getting though the tangle of *Hibiscus* and *Merremia* on the slopes proved even harder, until I reached the *Samoana* tree.

The view from the top

Mt. Turi from just below the peak I reached

That night the clouds allowed a good sunset and as I sat at the roulettes for supper the *Taporo VI* ferry pulled in to the harbour. This was the cargo ferry that I had taken to Raiatea 25 years ago. On that occasion it had also called in to Fare harbour on its way from Tahiti, but in the early hours of the morning, so I had seen nothing of Huahine beyond the lights of the harbour.

None of the early explorers gave any account of the forests of Huahine. Banks was uninformative as usual, but there ought to be more information from 1789. In that year the *Bounty* was preparing its breadfruit seedlings and the expedition took advantage of their enforced residence by exploring the islands more thoroughly than had been done before. At the end of the year they were on Huahine. As the longest European residence on Tahiti of the 18th century this expedition should have produced a wealth of information on the islands, and it is the most famous of all the expeditions to Tahiti. Unfortunately, this is for all the wrong reasons and in the subsequent drama Bligh's journal was lost. Of the two botanists, the assistant botanist William Brown went with the mutineers to Pitcairn island,

where he died in 1793. The chief botanist, David Nelson, chose to remain with Bligh and became the only casualty of Bligh's astonishing voyage across the Pacific Ocean. Nelson almost made it, dying in 1789 just after the launch in which they were marooned reached safety at Coupang. Subsequently Bligh wrote a narrative account of the voyage but, not unreasonably, this is entirely preoccupied with the mutiny and gives no useful information on the islands.

Bligh returned to finish his breadfruit mission in 1791-3, on HMS *Providence*. He was nothing if not single-minded and his journal simply records events. He was accompanied by George Tobin as ship's artist and his pictures are more informative.

With most of the day before my flight to the next island I had plenty of time to explore Huahine. Walking round the coast was very picturesque, with the views and the flower-filled gardens. Common noddy terns were nesting in a coconut tree and later I saw several slaty egrets and harriers.

The walk around Huahine-Nui is undeniably beautiful. It is not a place of palm-fringed beaches but mostly of palm-fringed rock

shores. Everything was very quiet and calm. The road-side houses were neat and tidy, the gardens a riot of flowers. It was all as a tropical island is supposed to be, as many used to be but so few are.

The south side of Huahine-Nui has a low agricultural area. I explored this for a while and then decided that I was not going to find anything else different and that it was time to return to Fare. Cutting in from the agricultural area I took what I thought to be a short-cut across the hills. Unfortunately it did not quite join up, and became a very long diversion. It did though give a view into the rural gardens inland, below the mountain peak. This was interesting, and very picturesque in a different, *Merremia* smothered way.

The road passed by the 'sacred eels of Faie'. All the guide books refer to these enormous blue-eyed eels and they are remarkable. They live in between the roots of mape trees growing along a river running though a village. Here they are fed, these days as a tourist spectacle, but they seem to have been looked after by the village for a long time. When I got to the spot there was a man standing in the river handing pieces of tinned tuna to a few tourists who were feeding the eels. There were about a dozen great eels varying from a metre to nearly two metres in length, most with startling pale blue eyes. They slowly writhed around and over one another, taking pieces of fish, all sinuous and languid.

The presence of the sacred 'puhi tari'a' are supposedly seen as signs of abundance and purity. Eels are particularly significant in Polynesia according to legend (everything has a legend, often involving an improbably large octopus or eel). The sun and the moon had a daughter who was to be married to the king of a lake, no reason for this seems to be necessary. With this distinguished parentage the princess Hina was radiantly beautiful (literally – she gave off flashes of light), but the king of the lake was a hideous giant eel. Although the sun and the moon didn't think this was a problem, Hina thought otherwise and fled from the arranged marriage, seeking protection from the god Maui. The eel king tracked her down but was captured by Maui and chopped into pieces. Being an eel he continued thrashing around, and also speaking, promising that a day would come when Hina would put her lips to his. With the logic of legend, Maui wrapped the head in banana leaves and told Hina to bury it in her village, but warned that if she put it down anywhere else the curse would come true.

Predictably Hina found the walk home hot and tiring, so she stopped to bathe, putting down her unpleasant burden. At this point the ground opened and swallowed the head, and from this spot a strange tree sprouted. It grew upwards like a gigantic eel. This was

the first coconut tree. The thirsty, and remarkably unperturbed Hina found that by putting her lips to the coconut (obviously after having made a hole in it first) she could slake her thirst. So it was a surprisingly benign curse in the end. Even so, I have to agree with the princess, the eels are far from attractive.

The last point of interest on Huahine was at the north-eastern end of the island. Between the mainland and the islands on the fringing reef were a series of fish traps. V-shaped lines of rock walls had been built into the lagoon forcing fish moving with the tide to concentrate in the points of the Vs. A narrow opening channelled the fish either into traps, or into spaces where they could be speared. Some had shelters built over this point. I'd not seen traps like these anywhere before, they seemed very ingenious and probably effective. On the near shore were remains of older history in the form of several marae from the past century or two. Huahine has a greater concentration of historical structures than any other island and has also been investigated archaeologically more extensively than elsewhere. Much of the information on the biology of the pre-European days, and even some pre-Polynesian, comes from remains of birds, shells and seeds from middens in these sites.

Shortly before I got back to Fare it started to pour and continued for the rest of the day. My flight was just after dark, and it was still pouring when I arrived on Raiatea 20 minutes later.

Raiatea

In 1992 I had travelled as cheaply as possible and had arrived on Raiatea on the *Taporo IV*. This cargo ship had sailed from Tahiti with not much cargo and around 20 passengers. The main cargo between the islands was clearly copra and the ship seemed completely saturated in coconut oil that had seeped out of the copra. After five minutes the smell was nauseating but at sea the breeze made it much more bearable.

It was a night sailing and quite pleasant for an hour, but once in the open ocean spray drifted across the deck and the crew insisted we all cram into a small airless room. By the time we arrived at Raiatea I felt very jaded and was determined never to use a copra ship again. The next time I flew.

On Raiatea I was staying in the town of Uturoa, in the unprepossessing Hotel Hinano Api. Despite an austere outside the rooms were very large, quiet and comfortable.

A particular interest with Raiatea was the rarely explored high altitude areas. In 2005 the botanist and head of research in Polynesia, Jean-Yves Meyer, had climbed to the top of Mt. Toomaru, the island's summit. There he had found a new *Partula* species, since named *Partula meyeri*. This site had been visited once before, by the botanist John Moore in 1927. Moore had collected several other *Partula* species which now seem to be extinct. No other biologists have been to the top. I hoped to get to Mt. Toomaru to see if *Partula meyeri* survived, and what else there might be there. Jean-Yves and I had planned that a helicopter would drop us on the summit and we would climb down after exploring the site thoroughly. At the last minute the helicopter company ceased operation and we were unable to find anyone willing to cut a path up the mountain (I later saw why). As a result the attempt on Mt. Toomaru had to be abandoned. Instead we would climb to the next highest, but much more practical, area of Temehani Ute Ute.

For the ascent to Temehani Ute Ute we had a guide, Pao. We picked him up from his house in a valley near the path which started from a water treatment station. It passed through dense escaped garden plants and into *Hibiscus* and mape woodland along a very scenic river, which we waded through three times. Oddly Polynesian paths never seem to be able to decide which side of a river they should be on.

As we started a gentle climb up the valley side Jean-Yves started to pick out more interesting plants. Animal life was also starting to become more interesting, with a stick insect and the nymphal skin of a cicada lying on the ground. The vegetation continued to improve along the ridge above the valley. As on Huahine there was a mixture of native trees and orchids broken up by patches of bracken fern.

At the end of the ridge was a steep slope of bracken, its steepness making it difficult to climb in places. This folded into a stream-bed, topped with a waterfall. The ascent here was a short climb up the rock face. Although the climb was not difficult I was pleased it was not raining.

From the top of the waterfall we were on the plateau of Temehani Ute Ute, in an amphitheatre where the water collected above the waterfall. It had taken us three hours to get there. Pao set off back down the path, saying he would return for us the next day.

Making our way up the valley

After a break for lunch we climbed up out of the amphitheatre and onto the plateau proper. Unlike Temehani Rai this is not a true plateau but a series of gentle slopes separated by streams or ditches cut by the rain. The vegetation was waist high in place, ankle high in

The waterfall (above) and Temehani Ute Ute (below)

others. There was a scattering of waist-high shrubs of *Pandanus temehaniensis*. In some places this formed thickets where they became small trees a little over head-high. There was some bracken fern and clubmosses bushes of white flowered *Alstonia*, patches of heather and clumps of sedges. The vegetation was dense enough and high enough to be quite hard to work through.

The weather was glorious and we had fantastic views of Tahaa, Bora Bora and, just visible in the haze on the horizon, Maupiti. We picked a camp site on the edge of the plateau, facing the view, perched above Raiatea's west coast. From there we could see the western edge of the plateau, running up to the steep sharp ridge of Mt. Toomaru.

Camp site view: the top edge of Temehani Ute Ute, with distant Mt. Toomaru

Once camp was set up we had a couple of hours of light left in which to explore further up the ridge towards the highest point of the plateau. We could not reach the high point itself as it was isolated from the plateau by a steep depression. We climbed down to the edge of this depression where a *Pandanus* thicket was rich with orchids and ferns. These trees had little animal life in them; there were some kite spiders and land hoppers, and a ground beetle but little else. The trees on the plateau had been more interesting, with a very specialised plant hopper living in them. At first glance these looked like small cockroaches, but this was due to their flattened shape, adapted to press between the leaf blades.

A *Pandanus* thicket and its botany – *Bulbophyllum* orchid

Back at the camp site we watched the slow sunset behind the island, listening to white-eyes in the forest below. Cloud whisps gathered behind us but the sunset remained largely clear.

Even after a long day a botanist's enthusiasm is boundless

Sunset over Bora Bora

Once the sun had set Tahitian petrels started up their mournful wailing all along the ridge. There was no sign of them during daylight, presumably they were nesting along the cliffs just below our camp. At around midnight a light rain started striking the thin canvas of the tarpaulin sheltering me. Although it was only light rain it trickled round the edges of the tarp and seeped upwards from the spongy ground. Once the rain started it carried without a break throughout the very soggy night.

Dawn was difficult to identify other than the switch from mournful petrel wailing to the joyous chorus of white-eyes. The light was dim and the nearby ridge could only just be made out in the thick cloud. Light but penetrating showers of cold rain came and went.

In these decidedly un-tropical conditions we explored the rest of the plateau using dead reckoning for where we remembered interesting valleys or patches of trees to be, and their dim shapes ahead. Walking was not bad, but instantly soaking with the sodden vegetation. On the east side of the plateau a cold wind and increasing rain made for very unpleasant conditions. Scrambling

A bleak dawn

down ravines yielded little of interest other than the first flowering apetahi bush. The apetahi is the emblem of Raiatea, *Sclerotheca raiateensis* being found only on Temehani where it has become rare through over-collecting, competition from invasive plants and damage by wild pigs. Not surprisingly its origins are surrounded by legends, all of which start out differently but end up in the same tragedy. It may have been a fisherman and his wife, a lovelorn maiden and oblivious warrior, or the king and his lover. The fishing couple have a row, the maiden is rejected or the king goes off to war. Whatever the cause, the woman climbs up to the Temehani plateau to contemplate her life. There she becomes suicidal and either slits her wrist (plausibly) of cuts off a hand and plants it in the ground (far less plausibly). Thus we have the white bloodless hand of the heart-broken woman.

The apetahi was first recognised by botanists from a specimen college by Jean Nicolas Eugène Vasco. Vasco was a naval surgeon and in 1847 was stationed in Polynesia on board the *Uranie*. He collected natural history specimens, with a particular interest in plants, brining 1,520 specimens back to Paris. Although Vesco did

The five-fingered hand of the apetahi

collect a great deal of material, conditions were not easy for explorers at this time. France had officially annexed the islands in 1843, and Queen Pōmare IV had fled Tahiti for sanctuary on Raiatea. Within months tensions between the French and Tahitians erupted into the Franco-Tahitian War. This conflict came to an end in 1846 when French forces defeated the Tahitians and the Queen accepted terms for Tahiti to become a protectorate of France, although Raiatea and the other Leeward islands held out until 1897.

Vesco's collecting in Raiatea was just a year after this conflict and at the least the independent Raiateans must have been suspicious of French visitors. On Tahiti itself, Vesco had been constrained; an acquaintance of his (supposedly the naval surgeon and botanist Jean Nadeaud although he is known to have been in Tahiti in 1956-9, *after* Vesco had left) described their plan to climb Mt. Marau. "In consequence of some rumours of native attacks, the authorities, when he [Vesco] applied the day before for permission, refused to allow him to expose himself. As I was in no danger I received a pass for myself and a native; however, the native was afraid to go, and so I was obliged to go by myself." He had heard rumours of the apetahi and was aware that it is an exception to a common rule that island flowers are small and inconspicuous. He commented "I have somewhere seen the observation that "the botany of islands is particularly interesting"; this may be the case, but I think it must be construed merely to mean that the study of the plants is interesting, for assuredly in general the plants of isolated islands are in themselves particularly uninteresting, so far as their mere beauty is concerned, and, for myself, I must confess that I always feel a sensation of fatigue at the idea of hunting out the name of a plant which does not recommend itself by beauty, utility, odour, or curiosity of structure."

After some time examining all the apetahi plants we could fine and searing for other points of interest we returned to the amphitheatre above the waterfall. Sliding down into the bowl of the amphitheatre we dropped below the blanket of clouds and there was the view of the coast once again. This was heartening but the view was quickly whipped away by the return of heavy rain. There was more water going over the waterfall but that made little difference. What had changed since our ascent though was that all the rocks were covered in flowing water; there were no dry rocks on which to to get a firm hand- or foot-hold.

We were pleased to see Pao when he returned at midday. We were even more pleased to see that he had brought a rope. Being able to climb down the slippery rock face with the use of the rope made descending the side of the waterfall considerably easier than I had feared. The rest of the descent was easy but by the end I felt that I had never been so wet or so muddy.

Descending the side of the waterfall

After two days of mountain exploration I restricted myself to exploring the town of Uturoa for a day

Uturoa from 1934 (University of Hawaii) and today

Uturoa and Mt. Tapioi in 1920 (above, photo: L. Gauthier) and now

Of Raiatea and Tahaa Cook wrote in 1869 "The land on Ulietea and Otaha is of a very hilly, broken and uneven surface - except what borders upon the Sea Coast - and high withall, yet the Hills look green and pleasent and are in many places Cloath'd with woods... The produce of these Islands and manners and customs of the natives are much the same as at King Georges Island, only as the Bread fruit Tree is here in not such plenty the Natives to supply that deficience, plant and cultivate a greater quantity of Plantains and Yamms of several sorts and these they have in the greatest perfection."

Parkinson struggled with the spelling of Raiatea and described 'Yoole-Etea' as "This island is, in many respects, much like Huaheine, and the country as much variegated; but this side of the island seems to have undergone some revolution; the inhabitants are but few, and poor, and have no political distinction of rank amongst them. The shagreen is in greater plenty here, and at Huaheine, than at Otaheite, where it was a searce commodity. They have also great plenty of taro, and eape. As to the bread-fruit it was but young; and of apples I saw none... This island [Raiatea] is but thinly inhabited, and some parts of it very barren."

The coast of Raiatea in 1773 (William : : 'View of Part of the Island of Ulietea')

Cook's expedition was based mainly around Opoa in the south-east of the island. Banks reported on **20 July 1769** "At noon today come to an anchor at *Ulhietea* in a bay Calld by the natives *Oapoa*, the entrance of which is very near a small Islet Calld *Owhattera*. Some Indians soon came on board expressing signs of fear, they were two Canoes each of which brought a woman, I suppose as a mark of confidence, and a pig as a present. To each of these ladies was given a spike nail and some beads with which they seemd much pleasd. .... On landing Tupia repeated the ceremony of praying as at Huahine after which an English Jack was set up on shore and Captn Cooke took possession of this and the other three Islands in sight viz. *Huahine Otahah* and *Bola Bola* for the use of his Britannick majesty. After this we walk together to a great *Marai* calld *Tapodeboatea* whatever that may signifie; it is different from those of Otahite being no more than walls about 8 feet high of Coral Stones (some of an immense size) filld up with smaller ones, the whole ornamented with many planks set upon their ends and carvd their whole lengh. In the neighbourhood

Morai or burying place from the *Endeavour* journal

of this we found the altar or *ewhatta* upon which lay the last sacrafice, a hog of about 80 pounds weight which had been put up there whole and very nicely roasted. Here were also 4 or 5 *Ewharre no Eatua* or god houses which were made to be carried on poles. One of these I examind by putting my hand into it: within was a parsel about 5 feet long and one thick wrappd up in matts, these I tore with my fingers till I came to a covering of mat made of platted Cocoa nut fibres which it was impossible to get through so I was obligd to desist, especialy as what I had already done gave much offence to our new freinds. From hence we went to an adjoining long house where among several things such as rolls of cloth &c. was standing a model of a Canoe about 3 feet long upon which were tied 8 under jaw bones of men. Tupia told us that it was the custom of these Islanders to cut off the Jaw bones of those who they had killd in war; these were he said the jaw bones of Ulhietea people but how they came here or why tied thus to a canoe we could not understand, we were therefore contented to conjecture that they were plac'd there as a trophy won back from the men of Bola Bola their mortal enemies. Night now came on apace but Dr Solander and myself walkd along shore a little way and saw an *Ewharre no Eatua* , the under part of which was lind with a row of Jaw bones which we were also told were those of Ulhietea men. We saw also Cocoa nut trees the stemms of which were hung round with nutts so that no part of them could be seen, these we were told were put there that they might dry a little and be prepard for making *poe*; we saw also a tree of *Ficus prolixa* in great perfection, the trunck or rather congeries of roots of which was 42 paces in circumference."

The area around Opoa was not very productive for the naturalists: **(23 July 1769)** "Dr Solander and myself go upon the hills in hopes of finding new plants but ill rewarded; return home at night having seen nothing worth mentioning." **(2 August 1769)** "Dr Solander and myself have spent this day ashore and been very agreably entertaind by the reception we have met with from the people, tho we were not fortunate enough to meet with one new plant." (4 August 1769) "Dr Solander and myself go upon the hills accompanied by several Indians, who carried us by excellent paths so high that we plainly saw the other side of the Island and the passage through which the ship went out of the reef between the Islets of *Opoorooroo* and *Tamou*. Our walk did not turn out very profitable as we found only two plants that we had not seen before."

Excursion to Tahaa

Strong winds blowing across the Raatea-Tahaa channel seem to be a frequent feature. When Cooks expedition visited the island on 28th July 1769 Banks described landing: "Wind still baffles us as much as ever. This morn hoisted out a boat and sent ashore on the Island of Otahah in which Dr Solander and myself took a passage. We went through a large breach in the reef situate between two Islands calld *Toahattu* and *Whennuaia* within which we found very spatious harbours, particularly in one bay which was at least 3 miles deep...

"The Island itself seemd more barren than Ulhietea tho much like it in produce, bread fruit being less plentyfull than Plantains and Cocoa nuts. The people perfectly the same, so much so that I did not observe one new custom or any thing Else among them worth mention; they were not very numerous but flockd from all Quarters

Tahaa from Uturoa harbour

to the boat wherever she went bringing with them whatever they had to sell."

Getting to Tahaa was more complicated than anticipated. The ferry departures between the islands seemed to bear little relation to the published timetables. The only one that seemed to operate was the *Te Haere Mauru*, but this claimed not to return on the same day. Nonetheless I took it to Tahaa; it seemed likely that there were other ferries not on the timetables, or a taxi boat, or something.

The ferry took us to a jetty on the south side of Tahaa and then round the east coast to Haamene. The island is very picturesque from the sea but its vegetation is clearly heavily invaded, with the tall invasive *Falcata moluccana* tree visible everywhere.

We docked by the entrance to Haamene bay rather than at the town as I had assumed. This was a 30 minute walk along the causeway fringing the bay. Inland of the causeway were mangroves and small houses, picturesquely different from the other islands.

Tahaa with the north-western point of Raiatea in the foreground and Bora Bora on the horizon in 1773 (William Hodges: 'View of the Islands of Otaha and Bola Bola with Part of the Island of Ulietea': National Maritime Museum, Greenwich, London)

Approaching Tahaa

The walk into Haamene

The path up the mountain was easy to find but had an unusual approach through cattle pasture. Behind the cattle a rough road wound through the forest. The vegetation was very lush but all invasive and there was little of interest other than a very fine butterfly. I had a very quick search of the forest and then returned to the coast to look for transport back to Raiatea.

In Haamene I found the *Tahaa Transport Rapide* ferry waiting to depart. So the return turned out to be very easy after all.

Heading up the mountain through cattle pasture, to *Merremia* jungle

Back on Raiatea

After the interlude on Tahaa I started my surveys of the lowlands of Raiatea. For this I would be working in two areas that I had studied in 1992, Hamoa and Faaroa valleys.

I stayed in Hamoa valley at the Pension Trois Cascades. This is a complex of small chalets in a smallish area of gardens at the start of the valley. It is run by two Frenchmen surrounded by their cats and dogs. The room was simple but comfortable and, being down a side road heading to farms and a few houses inland, it was very quiet.

It was strangely exciting to be back at one of my old field sites and straight away I went to see what it was like now. The valley has changed considerably; it's more open, with settlements on one side (and more open to the river on the other). As well as suburban clearance there is agricultural clearance and the old site where I had found the last surviving *Partula* snails in 1992 is an agro-industrial area. I could recognise it by the slope of breadfruit forest behind it and the row of *Gliricidia* trees along the track.

What I remember as a forest of breadfruit is now roadside invaded woodland. Looking under fallen banana stems I started finding flatworms; astonishing number of flatworms, and big ones at that.

After Haamoa valley I had a only a final site to visit: the great valley of Faaroa in the south of Raiatea. The long walk along the road to Faaroa was quite dull. Everything is more developed and less interesting than it was 25 years ago. Faaroa is still beautiful though. The buildings are vastly more sophisticated but not much more numerous.

I found the location of the old Pension Greenhill, now two very fancy residences. The road across Faaroa bay has been widened, the vegetation has been pushed further back and there are more houses. I went a little way on the cross-island road, having a memory that the old inland track crossed the river. This was clearly going in the wrong direction so I retraced my way and took one of the many agricultural tracks. This must have been very close to the route I used in 1992 but it was very different. There were patches of mape but it was mostly cleared and planted with coconuts (with cattle pasture underneath, as on Tahaa).

Faaroa is a mosaic of habitat: agricultural areas, bamboo thickets and mape forest. Although it is all very heavily invaded there are some interesting plants, like the *Malaxis* orchids.

Hamoa river

Faaroa bay and the river

In 1992 there was a distinct agricultural area and then a great expanse of bamboo forest. Exploring the valley I found I could make my way into the forest along clear paths. I assumed these would lead to other pockets of plantations but they became increasingly narrow until the ended in a jumble of bamboo stems. It was then that I realised that they were trails made by wild pigs. There was little of significance in the bamboo forest but old *Partula* shells could still be found on the ground. Bamboo forest is horrible habitat, having just the one plant species and few animals. Plus the stems are covered in minute irritating hairs. This forest was not too bad as the pig trail meant I could avoid physical contact with the bamboo much of the time. It was interesting that *Partula* had lived here; no-one had ever looked in the bamboo for interesting animals, and now these snails were all dead.

When the pig trail ran out I would have retracted my steps had I not been able to see different vegetation ahead. Carefully making my way through the bamboo I found myself in a narrow fringe of mape along the river headwaters. Climbing some rocks in the river I found a pocket of diverse vegetation, with *Freycinettia* and orchids. There were no animals visible but still it was interesting

to see that pockets of good habitat existed within the bamboo.

Now the bamboo itself has been extensively cleared and paths and roads cut across the valley floor, the mape forest patches seem more obvious. They are mostly along the river but some exist on the slopes as well. Even away from the river the trees have their fantastically sinuous buttress roots.

I searched pockets of different vegetation, finding an abundance of flatworms but also some reasonable snail diversity. There were areas of remarkably good vegetation right on the edges of new clearings. In here I found more different insects than I had seen anywhere else in the islands and a profusion of one species of jumping spider. *Thorelliola ensifera* is found throughout the Pacific islands and is known as the 'sword-bearing' jumping spider. This bizarre name refers to a spike projecting downwards at the front of the male's head. Given that the spider is only a few millimetres long it is not surprising that I didn't notice their 'swords'.

Unfortunately we have little idea of what Faaroa valley was like originally. The bamboo forest was certainly not the original vegetation, being a plant that takes over in cleared ground. Presumably Polynesian farmers cleared or burned much of the valley. This was probably before Banks spent time on Raiatea, but neither he nor any of the other early naturalists described what the place was like. We know Banks and Solander did see the bay but he had little to say about what he saw in the area, making unhelpful reports such as "Weather mended a little. Dr. Solander and myself go upon the hills in hopes of finding new plants but ill rewarded; return home at night having seen nothing worth mentioning."

The *Endeavour* initially anchored in Opoa in the south-east of the island but did sail north within the lagoon to Faaroa, there they sailed out of the lagoon with some difficulty: "Foul wind. The Captn attempts to go out of the reef at another passage situate between the two Islets of Opourourou and Taumou…

"Soon after this we came to an anchor and I went ashore but saw nothing but a small marai ornamented with 2 sticks about 5 feet long, each hung with Jaw bones as thick as possible and one having a skull stuck on its top."

The customs of the islanders, and the number of human skulls around the marae distracted Banks from paying attention to the habitats of the island.

Walking back from Faaroa bay the wide clearance of gardens and open spaces along the road afforded views back into the central mountains. I enjoyed being able to see the long plateau of Temehani Rai while standing dry in the warm sun.

The spectacular view of Mt. Toomaru (left) from Faaroa

Bora Bora

The flight from Raiatea to Bora Bora takes only 15 minutes and is very straightforward, although I was confused by the indirect route that detoured over Tahaa.

Until relatively recently Bora Bora had the only sizeable airstrip in French Polynesia (hence wet runways are not a problem, in contrast to neighbouring Maupiti). This dates back to 1942 when the U.S. military decided to place a refuelling base at Bora Bora. Initially this comprised lanes within the lagoon for use by seaplanes. At the end of 1942 work started on a runway on Motu Mute, suitable for servicing bombers and fighter aircraft. By 1944 the theatre of war had moved away from the south Pacific and in 1946 the air base was abandoned. Commercial flights started operating in 1958 and it remained the only airstrip in the islands until 1960 when Faa'a airport opened on Tahiti.

Bora Bora (or as he called it 'Bolabola') was described by Cook on 29th July 1769 as "This Island is very remarkable on account of a high craggy hill upon it, which terminates at top in two

U.S. air base in 1941, from Department of the Navy Bureau of Yards and Docks 1947

peeks the one higher than the other, this Hill is so perpendicular that it appears to be quite inaccessible." Parkinson commented "The island of Bolobola is made up of one very high forked peak of land, with seven low hills round it." It is an unlikely looking place in reality, more like an idea of an island than the real thing.

Banks' observations of Bora Bora are only slightly more detailed: "The wind last night has favour us a little so that we are this morn close under the Island of Bola Bola, whose high craggy peak seems on this side at least totaly inaccessible to men; round it is a large quantity of low land which seems very barren. Tupia tells us that between the shore and the mountain is a large salt lagoon, a

Bora Bora in 1849 by Fanshawe (National Maritime Museum, Greenwich, London), and today

certain sign of barrenness in this climate; he however tells us that there are upon the Island plenty of Hogs and fowls as well as the vegetables we have generaly met with."

As soon as we landed it was clear that this was a much more touristy island than elsewhere, with many hotel shuttle boats at the airport. The island shuttle boat was a large catamaran, a far cry from the little boat on neighbouring Maupiti.

The crossing goes along one side of the island, and no-one with a camera can resist crowding to the front or ones side of the catamaran. At Vaitape harbour I was met and driven to my accommodation at the Villa Rea Hanaa. This is an extremely odd place. I had chosen it for its view; being on the south peninsula it is one of the few places on the main island that give good views of the mountain. It is owned by and architect/artist and is delightfully eccentric: a Mediterranean style villa, painted bright orange and packed with eclectic painting and sculptures. It is impractically open plan and all a bit grubby, which felt in keeping.

I returned to Vaitape for supplies and a look around. It is a tiny town, dedicated largely to selling pearls to the tourists. Outside of the one street of the town there is hardly anything and the small houses are very ordinary. This is the most touristy and expensive of

the Society islands, but almost all the hotels are on the motus, the island is hardly touched, other than the pearl shops.

The mountain path runs up from Vaitape, firstly through an area of very poor housing. At the back of this is a banana plantation and then the steep slopes of the mountain forest. I started up what looked like a faint path (completely overlooking the real and obvious path to my right). This quickly turned into a climb up a slope of *Hibiscus*. Although steep, it wasn't hard going and very quickly I started finding old *Partula* shells. This was very surprising as they had been thought to have died out more than 20 years ago. Although the shells were all old it seemed unbelievable that they were that old.

Abandoning my stream-bed or pig trail, and cutting across the mountainside I reached an area of large mango trees. There I found the obvious path. Climbing up the slope I was passed by two noisy groups of local youth, clearly out for a Sunday morning excursion.

The climb was very steep but also easy, with the very clear path and ropes up for crossing rocks (although these weren't really necessary). I stopped at a rock face which proved interesting for

searching for snails, and then started back down the path. I did find more flatworms, including the bizarre hammerhead flatworm *Bipalium kewense* which hunts earthworms.

Hibiscus habitat on the mountain slopes

Widowed gecko (*Lepidodactylus lugubris*)

I had booked whale watching trip while on Bora Bora as my last opportunity to see them. Whales were around but the likelihood of finding them was far from certain due to the rather changeable weather. The excursion was well organised and put a great deal of effort into searching around the outside of the reef. Over four hours we covered quite a lot of water looking for whales, and saw nothing. The sea was rough, very rough on one side. Three times a hydrophone was lowered into the water, and on each occasion a male whale could be heard singing a very long way off.

The only notable things we saw were boobies, especially a feeding flock of mixed brown and blue-footed boobies. On the way back, at speed we put up a few flying fish, but mostly we saw waves.

My last excursion was a walk up the flanks of the northern side of Bora Bora's mountain. From below I had seen an open clearing and, as I thought, there was a road running up to it. After walking up a concrete road past large houses behind high walls, and cutting though tracks between shacks I came to the newly bulldozed clearing. Unlike on Raiatea where most clearing seems to be agricultural, this spot looked more like it had been picked for housing development. This building of large houses seems to be a distinctive feature of this

commerce-orientated island. Even so this is relative commercialism; most of the island is uninhabited and not everyone lives behind high walls. I'm sure the Bora Bora of today horrifies those who knew it in the past, but for a first time visitor it makes a very interesting few days stay.

Although the clearing made a scar on the mountain the very wide roadway provided sunning places for *Hypolimnas bolina*

Merremia covered trees and the flower of a *Hibiscus* poking through the creepers

butterflies, and for yet another harrier. The road had been there long enough for *Merremia* creepers to smother all nearby vegetation.

To one side of the clearing ran a dry stream bed. I dropped into this and used it to climb through the creepers and into the forest a bit higher up the mountain. Here were the familiar trees I had found on all the islands: mape and *Hibiscus*. This forest did not look particularly interesting and I thought it was probably secondary vegetation growing on much older clearings. Even so there was a surprisingly good variety of ferns growing in the river. Here again there were old *Partula* shells. Not as many as the day before but enough to show that in the not too distant part the snails had been abundant all around the mountain.

This brief scramble was my last excursion before starting the long journey back to Tahiti and to England. Although there was nothing outstanding about these last few hours, they did encapsulate the history of the forests of these islands. The new clearances for modern housing and agriculture cut into forests that are themselves growing on much older clearings. Over centuries Polynesians cleared patches of forest, exploited their natural resources and introduced all manner of plants and animals. The native fauna and flora that survive today have adapted to this changed environment, sometimes becoming closely tied to it. To some extent the few surviving *Partula* snails may depend upon the mape tree for survival, a tree brought to the islands by the Polynesian explorers.

Departing Bora Bora, view of the island and the fringing motus, with the resorts built into the lagoon

References

Banks, J. (ed. Hooker, J.D.) 1896. *Journal of the Right Hon. Sir Joseph Banks Bart., K.B., P.R.S. during Captain Cook's first voyage in H.M.S. Endeavour in 1768-71 to Tierra del Fuego, Otahite, New Zealand, Australia, the Dutch East India, etc.* Macmillan & Co., London

Banks, J. 1771. *The Endeavour Journal of Sir Joseph Banks 1768-1771.*

Bougainville, L.A. de 1771. *Voyage autour du monde par la frégate du roi "la Boudeuse" et la flûte "l'Étoile"; en 1766, 1767, 1768 & 1769*

Cassin, J. 1858. Mammalogy and Ornithology. In: *United States Exploring Expedition during the years 1838, 1839, 1840, 1841, 1842, under the command of Charles Wilkes, U.S.N.* Vol II. Lippincott & Co., Philadelphia

Christian, F.W. 1910. *Eastern Pacific Lands Tahiti and the Marquesas Islnds*, R. Scott, London.

Cook, J. 1893. *Captain Cook's Journal during his first voyage round the world made in H.M. bark "Endeavour" 1768-71. A literal transcription of the original MSS. With notes and introduction edited by Captain W.J.L. Wharton, R.N., F.R.S. Hydrographer of the Admiralty. Illustrated by Maps and Facsimiles.* Elliot Stock, London

Department of the Navy Bureau of Yards and Docks. 1947. *Building the Navy's Bases in World War II. History of the Bureau of Yards and Docks and the Civil Engineer Corps 1940-1946.* Vol. II. U.S. Government Print Office, Washington

Dierkens, M. & S. Charlat. 2009. Contribution à la connaissance des araignées des îles de la Société (Polynésie française). *Revue Arachnologique* 17(5): 63-81

Duperrey, L.-I. 18238 *Voyage autour du monde : exécuté par ordre du roi, sur la corvette de Sa Majesté, la Coquille, pendant les années 1822, 1823, 1824, et 1825.* Bertrand, Paris

Garrett, A. 1873. *Fische der Sudsee.* L. Friederichsen & Co., Hamburg

Lever, R.J.A.W. 1964. Whales and Whaling in the Western Pacific.

South Pacific Bulletin (1964): 33-36

Nadeaud, J. 1874. On the botany of Tahiti. *Trans. Proc. R. Soc. New Zealand* 6: 66-80 (anonymous when published)

Melville, H. 1847. *Omoo: A Narrative of Adventures in the South Seas*

Parkinson, S. 1784. *A journal of a voyage to the South Seas, in his Majesty's ship the Endeavour Faithfully transcribed from the papers of the late Sydney Parkinson, draughtsman to Joseph Banks, Esq. on his late expedition, with Dr. Solander, round the world.* Richardson & Urquhart, London

Thomas, W.S. 1979. A biography of Andrew Garrett, early narualist of Polynesia: Part 1. *The Nautilus* 93: 15-28

Wallis, S. 1773. *Voyages in the Southern Hemisphere*, vol. 1. Strahan & Cadell, London

Wilkes, C. 1845. *Narrative of the United States Exploring Expedition During the Years 1838, 1839, 1840, 1841, 1842.* Vol. 2 G.P. Putnam

www.ingramcontent.com/pod-product-compliance
Ingram Content Group UK Ltd.
Pitfield, Milton Keynes, MK11 3LW, UK
UKHW021956220326
11408UKWH00003B/343